CAN THE WHALES
BE SAVED?

CAN THE WHALES BE SAVED?

Questions about the natural world and the threats to its survival answered by the Natural History Museum
Dr Philip Whitfield

SCHOLASTIC INC.

New York Toronto London Auckland Sydney

Copyright © 1989 by Marshall Editions Limited.
All rights reserved. Published by Scholastic Inc., 555 Broadway, New York, NY 10012, by arrangement with Viking Penguin Inc.
Printed in the U.S.A.
This book was conceived, edited and designed by Marshall Editions Limited.
Editor: Carole McGlynn.
Art Editor: Daphne Mattingly.
Editorial Research: Jazz Wilson.
Picture Editor: Zilda Tandy.
Managing Editor: Ruth Binney.
ISBN 0-590-48663-2

1 2 3 4 5 6 7 8 9 10   14   01 00 99 98 97 96 95 94

Contents

Introduction

All over our planet animals and plants are in danger. From the mighty whales that swim in the oceans, relentlessly pursued by human hunters, to the vast expanses of the rain forests, chopped down to make room for agriculture, the future of wildlife is under constant threat.

As you look at the beautiful world around you, which you share with so many other living things, you may wonder why nature seems so complicated, and why it matters so much to care for wildlife. Why, for instance, are there so many different sorts of plants and animals? How do they all manage to live together, and which ones are most in danger of dying out? Do living things ever help each other? Can forests recover from forest fires?

The answers to these and many other questions show that living things exist together in lots of different ways, and depend on one another to "give and take." Through these sorts of relationships, whole groups or communities are built up. On land, these may take the form of forests or grasslands, for example. In the water, communities of plants and animals live in ponds, rivers, and oceans.

Human beings are also a part of the natural world, but the five billion people that make up the human race are putting terrible pressures on the Earth's living framework. We are destroying the forests, creating deserts and dust bowls, polluting the seas, making animals extinct, damaging the atmosphere, and changing the climate.

The future of our wonderful world depends on us, and the way we behave in the future. So we must look for ways in which harm can be avoided or reduced. If you have ever worried about the threats to the natural world, or wondered what can be done to safeguard our planet, this book gives some positive answers. It explains, too, the harmfulness of acid rain, the importance of the ozone layer, and the possible impact of the greenhouse effect.

From the youngest to the oldest, there is much that each one of us can do to help secure the world and its wildlife for the generations that are to come. The pages that follow give a lead to just some of those actions.

Why are some animals better at surviving than others?

Whether animals are good at surviving in the wild is partly a matter of chance, but is also linked to the way that different animal species adapt to their surroundings, and to the rate at which animals breed to create new offspring. Most important of all is breeding. If a creature gives birth to the young as fast as or more quickly than animals are dying, then the species will survive.

The chance element of survival is usually a matter of being in the right place at the right time. If an animal has the bad luck to live in an area where a sudden change in climate or natural forces makes survival impossible—due, for instance, to freezing cold or the eruption of a volcano—then no amount of adaptability or rapid breeding will enable it to prosper.

Changes produced by people can also affect an animal's chances of survival. Humans may destroy or alter habitats or nesting sites, remove animals' foods, or kill animals for pleasure or because they are pests—changes which could wipe out an entire species.

Adaptability is all about being flexible. Animals that can eat only one sort of

The wolf is a good example of an adaptable animal whose breeding rate is successful, yet whose survival is threatened by the activity of humans. Wolves are intelligent wild dogs that can live in different types of country, from open tundra to deep forest, and feed on a variety of prey animals.

But because wolves have been killed off by farmers over the last few centuries, they are now almost extinct in Western Europe. In less-densely populated areas (such as the tundra of the USSR), or where they are protected, wolves continue to flourish.

food, live in one sort of tree, or only flourish in a narrow range of temperatures are in perpetual danger. If the habitat changes, they will not be flexible enough to survive. On the other hand, animals that eat a wide variety of foods and live in different kinds of habitat and climate have a kind of built-in "insurance policy." Even if their surroundings change, these flexible animals can change their lifestyle and carry on.

What are the most successful plants?

Successful plants come in a wide variety of shapes and sizes, some small and some very large. Some plants grow, set seed, and die in just a few weeks, while others live for literally thousands of years.

Plants have a whole range of different strategies for survival. At opposite ends of this range are tiny garden weeds like chickweed and huge forest trees such as the redwoods. These very different plants are extremely successful in contrasting ways.

Weeds such as chickweed and shepherd's purse have a strategy of reproducing as quickly as possible once the conditions for growth are just right. Waiting as minute seeds in the ground, they can grow in just a few weeks to a full flowering size as soon as sufficient light, water and heat are available. In this short time they can produce a new crop of tiny seeds to await the next growth opportunity.

At the other extreme of plant life survival tactics are the giant forest trees. Their technique is one of slow growth, a very long life and a large size. Although there are several years before seed production starts, the seeds are large and usually stuffed with food reserves to ensure a good start for the new seedlings, and are produced regularly—usually once a year.

An oak tree, for instance, may not start producing flowers until it is ten years old or more. From this time on, though, it may live for over 500 years and grow to nearly 100 feet (30 meters) in height. Acorns are made each year; every cup contains a large, food-packed seed.

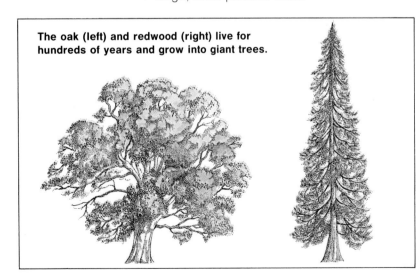

The oak (left) and redwood (right) live for hundreds of years and grow into giant trees.

3

What food do plants need?

Unlike animals, plants do not need to "eat" food to survive and grow. Plants are made of the same types of substances as animals—proteins, fats, and carbohydrates such as sugars and starches. But plants are much more independent than animals. All green plants have the magical ability to make these complicated materials from scratch. Because of this, plants are often known as "primary producers."

Plants start off with the simplest possible ingredients—water, mineral salts from the soil, and the gas called carbon dioxide from the air. The green parts of plants trap energy from the sun and use it to change these starting materials into all types of new materials or "building blocks." The plant then uses these new materials to keep itself alive and construct new, living cells in its roots, stems, leaves, and flowers. The light-powered manufacture of new living matter is called photosynthesis—that is, "light-building."

Sometimes, plants cannot easily make food by photosynthesis. To survive through these lean times they store foods like starches, fats, and proteins in their cells. These stores can, when needed, be used to provide energy or for growth.

One example of this is the food reserves in seeds. For seeds to germinate underground they need energy. They also need to grow a little before they get into the sunlight and make leaves which can photosynthesize. When we eat seeds such as beans, peas, and cereals, we are eating the seeds' nutritious food reserves.

4

Do all plants need sunlight?

All green plants need sunlight. The light energy that streams from the sun is trapped by the green parts of plants and enables photosynthesis to happen, so that the plant can manufacture the materials it needs to survive and grow. Plant leaves are light traps: they often grow face-on to the sun, to catch as much light as they can.

The green of the traps is due to a special colored substance called chlorophyll. The chlorophyll is gathered together inside plant cells in tiny green structures called chloroplasts. They absorb sunlight energy and then pass it on through a chemical chain.

This chain uses the trapped energy for two processes. First, it makes a constant supply of an energy-rich substance which activates the living processes that go on in the plant's cells. Secondly, it changes carbon dioxide from the air, and water from the soil, into sugars. This is the crucial first step of photosynthesis.

Only a few plants can live without sunlight. Some parasitic plants, for example, have little or no chlorophyll and are white or cream. They attach themselves to ordinary green plants which *can* photosynthesize. Using rootlike organs, the parasitic plant sucks the sugars and other foods from the sap of the green plant, stealing its trapped sunlight energy.

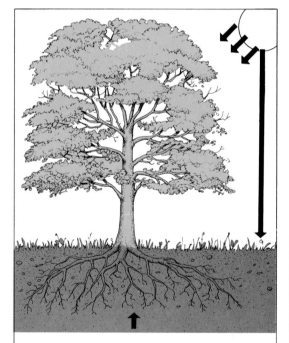

The tree above shows the way green plants use photosynthesis to change new materials into the substances they need for life and growth. The chlorophyll in the plant leaves traps sunlight energy while carbon dioxide from the air gets into the plant through tiny pores in the leaves. Water from the soil is drawn up into the tree by its roots. It is pulled and sometimes pushed up to the leaves through the trunk. Mineral salts such as nitrates, phosphates, and sulphates from the soil are dissolved in water and help to build new living matter.

How do plants grow?

Most plants grow at their growing points, the stem tips or root tips. We are probably more used to the growth of babies or other young animals, where an "all over" type of growth takes place, like a balloon slowly being blown up. By concentrating their growth at the growing points, flowering plants grow in a different way.

The position of a plant's growing points, or stem tips (usually found at the centers of buds), decides the overall shape of the plant. If one topmost bud is the main growth point, the plant is a tall, thin one. If growth also occurs from side buds, the plant has a bushier shape.

In a mirror-image of the above-ground plant, the underground branching of the root system also depends on the number of growing root tips. A plant such as a carrot with one main central tip has a strong, deep, single tap root. Where there are several side tips, the root system spreads sideways in the soil.

Even when all the leaves fall from a broad-leaved tree in winter, the growth tips, though inactive, are still inside the well-protected buds. In the spring they "come to life" and re-start their growth.

Each growth tip is made of a cluster of rapidly dividing cells known as meristem cells. These manufacture the new cells that plants need if they are to grow. In woody plants, other meristem cells are found in a layer between the bark and the wood. This layer produces new bark cells and new inner trunk cells, so that tree trunks can grow thicker as well as taller as the tree ages.

The leaf buds of a horse chestnut tree burst open in spring as growth starts anew.

6

Why do some animals eat meat and others plants?

All animals have to eat some kind of food. They almost always eat plants or the bodies of other animals which contain the complex substances they need. Unlike plants, which can manufacture food from simple raw materials, animals have to "steal" what they need in ready-made form from plants or from other animals which have, in turn, eaten plants.

Animals are often described by what they eat. Carnivores eat only other animals. Herbivores eat only plants or parts of plants. Omnivores eat both animals and plants.

Examples of carnivores include eagles, dragonflies, sharks, cats, dogs, and crocodiles. They tend to be fast, powerful animals with jaws, teeth, claws, or beaks adapted for catching and killing their prey.

Cows, deer, elephants, termites, and locusts are all herbivores, whose diets consist of plants alone. Their mouths and guts are specialized for grinding up and digesting tough, fibrous plant material (see Question 36).

Humans, and also chimpanzees, bears, and rats, are omnivores. We can eat a mixed diet of animal and plant material.

Two lionesses from a family group called a pride, hunt together to bring down and kill their prey. Lions are big cats, and excellent examples of large carnivorous animals. Their diet is the bodies of the other animals that they kill, which are mostly grass-eating herbivores such as the wildebeest.

A food pyramid shows "who eats whom." In this example of the ocean, the pyramid is turned on its side. The "base" of the pyramid is inhabited by the primary producers, minute plants called algae which make food from simple materials by photosynthesis. These are eaten by the slightly bigger animal plankton. Small fish feed on the animal plankton and are eaten by larger fish. Toward the "top" of the pyramid, a few big carnivores such as tuna and seals eat the larger fish. Some top predators such as a killer whale consume the seals. Humans are the only enemies of most top predators.

If animals eat each other, why don't more die out?

It is easy to imagine that because lions kill wildebeest, owls eat mice, and tuna catch mackerel, that wildebeest, mice, and mackerel are in danger of becoming extinct because they are the hunters' prey. In almost all cases, though, the death of some animals through being hunted and killed by carnivores is simply the loss side of a budget of survival—along with deaths due to disease, old age, and starvation.

When animals in a particular area are counted, year by year, the usual pattern is one of more or less unchanging numbers. Most populations are, in fact, stable in size. For this to be so, the losses due to accidents, disease, or killings by other animals must be almost exactly canceled out by the births of new animals. Reproduction is the gain side of the budget, and for a species to survive in its natural habitat it must be able to breed fast enough to make up for normal losses.

This means that animals which suffer a high rate of deaths are those that also have very high birth rates. At the other extreme, animal species which breed at a slow rate are the ones that must ensure that they lose few of their offspring. A lion cub, for example, is tended by the pride for up to 18 months after it is born.

These beautifully balanced natural systems can be severely damaged by humans. By destroying environments or killing animals we can stop natural breeding rates from being able to keep up with losses. If this happens for long enough, a species may die out.

Cod produce millions of eggs to offset huge losses

Frog spawn contains a few thousand eggs

A mouse can produce up to 100 offspring a year

A cuckoo, with no nest to build, can lay 20 eggs in a season

A blackbird produces 4 to 5 young birds in a clutch

An elephant is pregnant for a year and produces only one calf

8

Can plants kill each other?

Yes they can. Although we normally associate this more typically with animals, plants *are* capable of killing each other.

A plant may kill another plant almost accidentally. This happens when plants compete with each other for essential life-giving resources. A good example is the competition for light. In a forest, rapidly growing trees with dense foliage can so completely cut off the supply of sunlight to plants underneath them that the lower plants are killed. In a beech woodland, or some coniferous forests, almost no plants grow in the soil in the shade of the trees.

Other types of plant killing are more direct. Some are killed by true parasitic plants, which draw food materials from another plant instead of making their own. The tropical parasitic plant called witchweed is a serious pest of cereal crops like millet and sorghum in Africa. The witchweed sucks water and nutrients out of the cereal until, in most cases, the host plant wilts and dies.

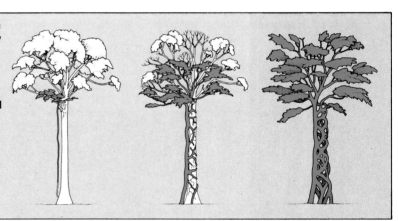

The strangler figs found in tropical rain forests slowly smother the forest trees on which they grow. The strangler climbs up the tree trunk, spreading upward and sideways until it surrounds the trunk in a cage of thick stems. The foliage from these stems overwhelms that of the host tree, which eventually dies.

9

Can plants kill animals?

Indeed they can. Around the world in both temperate and tropical regions there are hundreds of plants that can kill animals. They are known as carnivorous plants.

Most of the best-known carnivorous plants grow in soils that are poor in vital mineral salts like nitrates and phosphates. Examples include the Venus's flytrap of eastern North America, the sundews of marshes and swamps, and the pitcher plants of tropical rain forests. They all catch and digest small animals in order to extract from their bodies the salts missing from the soil.

The prey animals are usually insects and other small invertebrates. Although some of the larger pitcher plants can sometimes kill a small mouse, there are certainly no plants that can kill people!

Besides carnivorous plants that kill for food, many other plants have the ability to poison and sometimes kill animals that try to eat them. Such plants produce a range of toxins as a kind of self-defense, to protect themselves from herbivores.

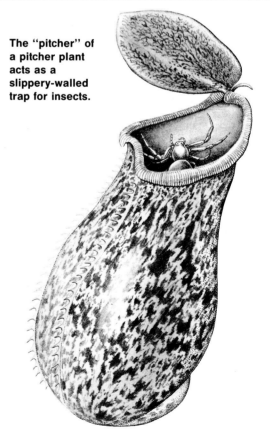

The "pitcher" of a pitcher plant acts as a slippery-walled trap for insects.

How do they do this?

The carnivorous plants have developed an extraordinary range of ways of trapping and digesting animals. The traps they make are usually formed from specialized leaves. They work either by being a sticky surface such as flypaper, as a pit into which the animal falls, or as an active trap which grabs the prey.

The sundews are sticky trappers. Each plant has a rosette of round leaves at its base and each leaf is covered with glistening, sticky-topped hairs. A fly, ant, spider, or beetle which lands on a leaf or crawls across it becomes trapped in the glue on the hairs. The hairs bend over and the leaf curls up to enclose and digest the trapped animal.

The pitcher plant traps are shaped like smooth-sided vases with a pool of digestive fluid at the bottom. Attracted by sweet secretions at the top of the vase, an insect easily loses its footing and falls into the pool to be digested.

The Venus's flytrap is an active trapper. The ends of the leaves have edges fringed with spines. When a fly lands on the leaf, sensitive hairs are stimulated and cause the edges of the leaf to close over the fly, making a cage. Inside this, digestive juices are produced to break the insect down.

There is even a water plant that traps prey underwater. This is the bladderwort. On the underwater leaves are tiny chambers, or bladders. When a water flea or other small invertebrate swims near the mouth of a bladder, the bladder suddenly sucks in water, pulling the animal in with the flow.

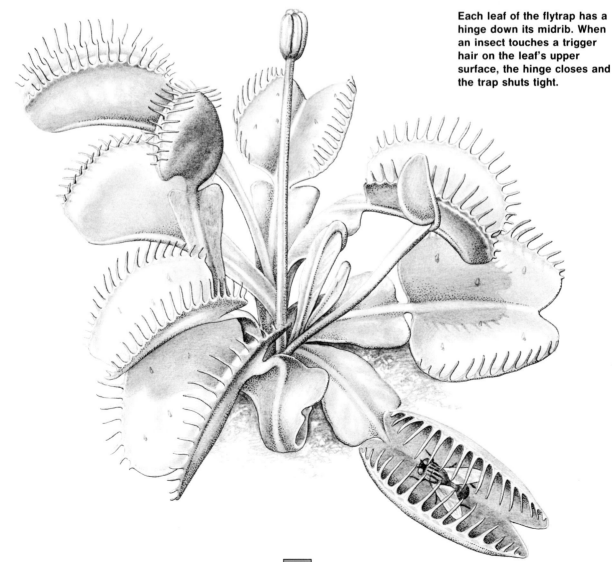

Each leaf of the flytrap has a hinge down its midrib. When an insect touches a trigger hair on the leaf's upper surface, the hinge closes and the trap shuts tight.

11

Where do new types of animals come from?

New types of animals—that is to say, new species—are produced by changes in their genetic blueprints—their inherited plans for growth.

Although they look very similar, every individual animal in a species has slightly different genes from all the others. These make each animal unique in hundreds of small ways. Imagine that the differences mean that an animal gets more food than its neighbors. If this allows the creature to survive better, and have more young than others in the same species, its genes will be more common in the next generation,

and the species will have changed slightly. These small changes add up over the generations to create bigger alterations.

Completely new species also develop when some kind of natural force—such as the sea, an ice sheet, or the creation of a new mountain range—splits a species into two or more groups. Once separated, the animals will change in different ways.

If the groups eventually come back into contact again, they may have altered in such a way that they cannot successfully breed with each other. A single species will have been split into two or more species.

The hooded crow is a two-tone bird in black and gray. It lives in Scotland, Scandinavia, and Eastern Europe.

The carrion crow is black all over. It lives in England, Ireland, and Western Europe.

In the last Ice Age the advancing ice sheets pushed the crows' ancestors south. They split into two groups, one on either side of the Alps.

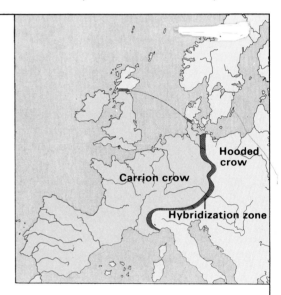

The groups adapted differently, and when the ice retreated, there were two crow types which could breed only in a "hybridization" zone.

12

How are new types of plants created?

Most new plant species are produced in the same ways as new animal types. But although the cross-breeding of two distinct species is usually impossible, sometimes it happens by accident and produces a new, successful type of plant.

Plant breeders can create new plant varieties, but these are only the different kinds of plant within a species. They do this by selective breeding—choosing the parent plants with great care in each generation. In this way they can breed new shapes, sizes, or colors of flowers, or heavier-cropping food plants. The huge head of a modern farm variety of wheat has been bred from the tiny head of the wild version.

Wild wheat   **Farm wheat**

Are new sorts of animals still being discovered?

Yes, all the time. Zoologists and explorers all over the world are constantly discovering and naming new species.

Since most parts of the planet have now been carefully explored, it is rare for new big animals to be found—it is difficult for a large animal or bird to stay unnoticed for long! Such animals turn up only in inaccessible places, such as the beautiful okapi shown here, which was discovered in the forests of Zaire in Africa as recently as the beginning of this century.

New species of small invertebrate animals, particularly insects, are found in the hundreds every year, and there are still plenty more to be discovered. This is because there are hundreds of thousands of these tiny animals in existence. They are also hard to tell apart, and there are fewer specialists studying them.

The okapi, a member of the giraffe family, was only discovered in 1901.

NEW ARRIVALS
In a single issue of the *Journal of Zoology* in 1988, new monkey, fish, and sea cucumber species were described for the first time. The sun-tailed monkey, with a bright orange tip to its tail, was found in the jungles of central Gabon in West Africa. The fish was a brightly striped, short-lived species from a lagoon in Brazil. The sea cucumber was found on a sea mount on the floor of the southeastern Atlantic Ocean.

14

Why does the world have so many different plants and animals?

There are almost certainly over a million species of living things in the world—plants, animals, and microbes. Some biologists have recently suggested that ultimately there could be as many as 30 million to be discovered.

The number is so high because each kind of plant or animal—that is, each species—has its own special lifestyle or "niche." (A niche is a total description of the way an organism lives—exactly where and exactly how.) Because there are millions of possible niches in the world, there are similar numbers of species.

The reason the number of niches is so great is because climate, terrain, and opportunities vary so much from place to place. In addition, the vegetation types that flourish in different regions produce a whole new set of niche possibilities. The plants provide food, shelter, and nest sites for many of the animals and microbes that live in the same area.

In a similar way, every animal and plant is a "home" to other creatures that live in or on it. Some of these are helpful partners (symbiotes), others are damaging (parasites and germs). All these "hangers on" are using a new set of niches based on the bodies of living things. So, for instance, one species of animal will have its own special set of parasitic worms and microbes that live on that species alone. These sorts of lifestyles multiply the number of niches available even further.

With all these possibilities for different ways of life, there seems to be almost no limit to the number and variety of living things that the Earth can support.

15

Can new ones be made artificially?

In a way, yes. Scientists are now able to tinker with the growth of creatures from the time they are single fertilized eggs. In addition, in the process called genetic engineering, they can put genes from one organism into the cells of another, and put laboratory-made genes into existing organisms.

By mixing early embryo cells of the two animals, it has been possible to make an animal that is part-sheep/part-goat (people are not sure whether to call it a shoat or a geep!).

Bacteria with new genes in them can be used to make complicated and useful drugs such as human insulin for diabetics, growth hormone for children with growth problems, and a drug called interferon which may help cure some forms of cancer.

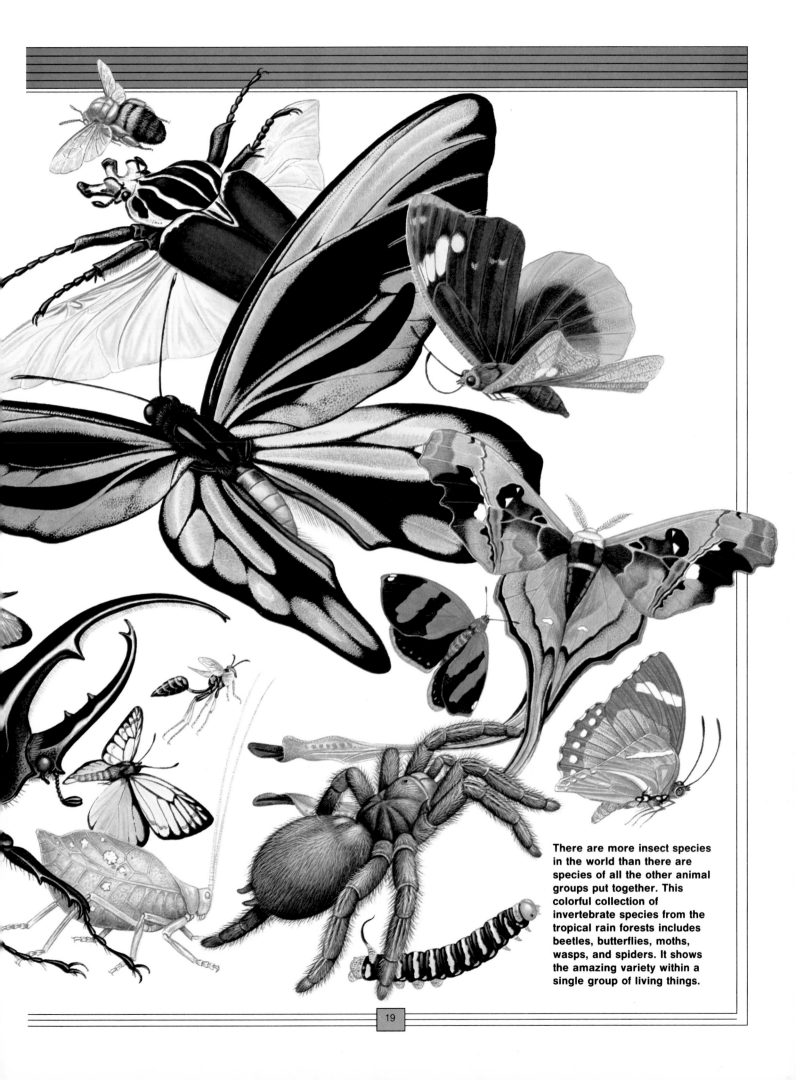

There are more insect species in the world than there are species of all the other animal groups put together. This colorful collection of invertebrate species from the tropical rain forests includes beetles, butterflies, moths, wasps, and spiders. It shows the amazing variety within a single group of living things.

16

Do the same types of creatures live every-where?

The answer to this question is no. Most animals and plants have a particular part of the world to which they are restricted. This may be the one where they first evolved and to which they are especially well adapted, or an area into which they have moved.

Even if there are other regions where these creatures could exist successfully, there are often barriers to travel which make it difficult or impossible for an animal or plant species to reach them. Wide stretches of ocean prevent the spread of a species of land animal that cannot fly. Cold zones can hinder the spread of plants or animals better suited to warmer weather. The snowy tops of a mountain range present an insurmountable barrier to an insect that must live in the hot, humid tropics.

Because most living things are fussy about where and how they live, we can describe a typical mix of animals and plant species for each region of the world. Many of the types in this mix will be found only in that region.

Polar bears, for instance, are found only in the snows of the Arctic region. They do not exist in the similar conditions of the Antarctic (see Question 45). Most species of penguins are restricted, by contrast, to Antarctica and the seas and islands surrounding it.

Hot climate areas give similar stories of restrictions in the spread of plants and animals. Today, Africa is the only place in which to find large numbers of giraffes, lions, chimpanzees, gorillas, cheetahs, and aardvarks. Although similar conditions occur in South America, none of these species are found there. Cacti, on the other hand, are found only in the dry lands of the Americas—none grow naturally in the deserts of Africa, Asia, or Australia. However, one forest cactus does grow in Africa.

As always with any rule of biology, there are some exceptions. Today, a few species exist that have an amazingly wide distribution around our planet. Birds, with their ability to fly great distances, provide some of the best examples. Both the barn owl and the fish hawk (also known as the osprey) are found almost worldwide.

The map shows the seven major regions of the world and their special animals.

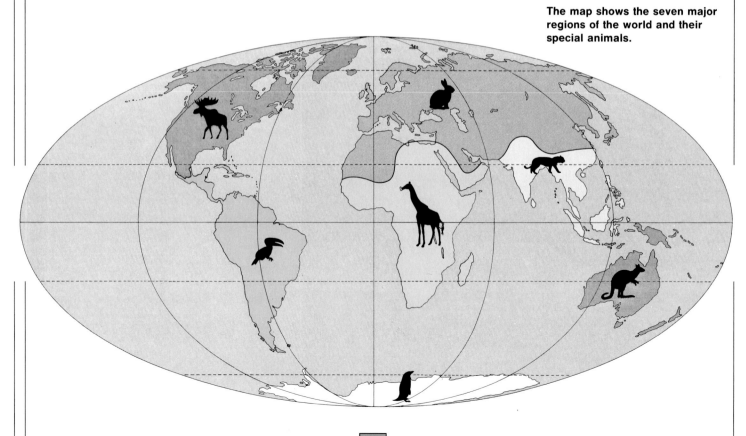

17 ⊟

What is an ecosystem?

An ecosystem is a characteristic mixture of plant and animal types that occurs in a particular landscape and climate. Another technical word for it is biome.

For an ecologist—a scientist who studies where and how living things exist—it is the plant life that provides the important framework for the ecosystem. The patterns of plant growth and plant types lay down the basis of any ecosystem. Its animals and microbes fit into the places or niches that the plants provide. And the plants, since they can use photosynthesis (see Question 3), provide all the new living matter in an ecosystem. The animals in the ecosystem depend on the plants for their food.

The plant framework of an ecosystem is decided by climate: the rainfall pattern and the temperature month by month. About ten major types of land ecosystem are known. These are mountains, tundra, cold coniferous forest, temperate forest, temperate grasslands, temperate rain forest, tropical rain forest, savanna, scrubland, and desert. Each one has a special pattern of yearly temperature and rainfall associated with it. In addition, the oceans, lakes, ponds, and rivers provide different aquatic ecosystems.

To take the example of a particular land ecosystem, the savanna landscapes, such as those found in East Africa, have a high temperature all through the year. The rainfall, though, is seasonal, with extremely dry and extremely wet seasons alternating. In these conditions typical savanna vegetation develops, with scattered trees and scrub in tall grass.

In contrast, temperate broad-leaved, deciduous forest, the type of habitat that is found commonly in the eastern states of North America and in much of Europe, has a much more seasonal temperature pattern. It drops to an average of 40° Fahrenheit (5° Celsius) or less in the winter, and reaches about 77° Fahrenheit (25° Celsius) in the summer months. There is more rain than in the savanna, and it falls throughout the year.

These climate differences are enough to create the dramatic contrasts between the savanna landscapes of Kenya and the oak woodlands of North America or Europe.

Tundra

Coniferous forest

Deciduous forest

Rain forest

Savanna

Desert

Swamp

Ocean

How many types of forest are there?

Most plant ecologists—scientists who study where and how plants grow—agree that there are three or four main types of forest. A forest is an area dominated by closely growing trees, and the type of forest depends on the climate—the range of temperatures and the amount of rainfall—in the particular area. The principal types are: cold coniferous forests, temperate broad-leaved forests, and rain forests. The rain forests are split into temperate and tropical sorts by some experts.

The cold coniferous forests, also called boreal forests, consist of tall evergreens—trees that have a covering of leaves all year round—like pines, firs, spruces, and larches. These trees have narrow, needle-like leaves, and cones to protect their seeds; they usually have a tall, pointed, tapering shape—like a Christmas tree.

Beneath the main trees are found evergreen dwarf shrubs. Coniferous forests grow where the average yearly temperature is below freezing point and where it rains little, and only during the short spring and summer periods.

The temperate broad-leaved forests are found in warmer zones where there is some rainfall throughout the year. These forests consist mainly of a mix of deciduous trees, whose leaves fall in autumn and winter, such as oaks, beeches, elms, sycamores, maples, and birches. The large trees form a broad leafy "roof", or canopy, beneath which are shrubs and then a lower, ground-covering "herbaceous" layer.

The rain forests, often called jungles, are found in the tropics and subtropics where there is a lot of rain (see Question 21).

The three photographs show the typical appearances of three forest types around the world: the broad, spreading treetops of a temperate, broad-leaved woodland (top left); the snowy landscape and dominant tree shape of a cold coniferous forest (left); and the luxuriant plant life of a tropical rain forest (above).

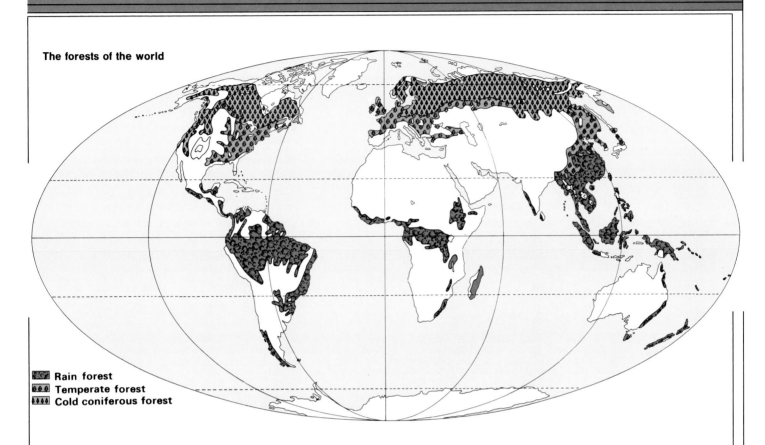

Rain forest
Temperate forest
Cold coniferous forest

19

Are they found in different parts of the world?

Yes they are. The main forest types, like all land ecosystems, are adapted to particular yearly patterns of temperature and rainfall. So the different climate belts around the planet provide zones suitable for the growth of contrasting sorts of forest. The map shows how the main forest types are spread in irregular stripes around the world; these correspond to climate zones.

Almost all the world's cold coniferous forests are in the northern half of the planet. They form a kind of ring stretching around the world. It starts as an unbroken stripe across North America, from Alaska in the west to Newfoundland in the east.

In the old world they are found in Scotland and Scandinavia, then in a broad band right across the USSR.

By contrast, the tropical rain forests of the world occur in three major, separate clumps where climate conditions are right. These are in South America, centered on the Amazon region, in West and Central Africa, and in Southeast Asia, including the islands of Indonesia and New Guinea.

Deciduous forests grow in temperate regions of the world, those with marked seasonal differences—cold winters and warm, wet summers. They are found in the middle latitudes of Europe, the USSR, eastern North America, and eastern Asia.

20

Can a forest recover from a forest fire?

Some forests are able to recover very well from fires. In dry areas of the world, forest fires are a serious risk—due, for example, to lightning strikes. But these forest trees may have evolved the ability to survive and regrow after serious fire damage.

The gum or eucalyptus forests in some parts of Australia are a good example of this. The foliage, bark, and the leaves that cover the forest floor are so dry that they burn rapidly in a fire, but some well-

protected buds can be left intact. From these, new growth can spring after the fire has passed. Some trees and smaller plants in this type of forest even have seeds and fruits which actually need to be damaged by fire before they can germinate into new plants.

Even after very serious fires, some living seeds are left in the soil, where they have been protected from the heat of the flames. From these seeds new seedlings grow.

What is a rain forest?

Rain forests are a type of woodland that grows in hot, wet parts of the world. A tropical rain forest is the most luxuriant and complicated sort of woodland on Earth. Rain forest in fact contains far more species of plants and animals than any other ecosystem on the land. An area of rain forest measuring only 4 square miles (1,000 hectares) may hold 750 tree species, more than 1,000 types of smaller flowering plants, 400 bird species, 125 mammal types, and 150 species of moths and butterflies.

Besides the variety, the density of living things in a rain forest is also enormous. Plants and animals breed and grow quickly because of the climate.

The rain forest landscape is one of almost unbroken tree cover. Over one and a half million square miles (400 million hectares) of our planet's surface is covered in this way. Most of the rain forests are in South America, Africa, and Southeast Asia, but there are smaller zones in Central America, the Caribbean, Madagascar, India, and Australia. The tree species, and the animals that inhabit them, differ in each area.

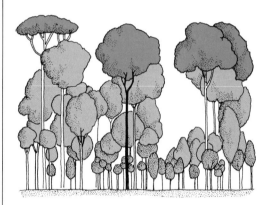

The varying heights of the different tree species and the other plants in the rain forest make up a multi-storeyed "building" of plant life. The diagrams above show two different jungle profiles.

The illustration (right) shows the rain forest levels in a more realistic way. At the top (1) are the "emergent" trees, some over 150 feet (55 meters) high, whose tops stick up above the main layer beneath. Next is the canopy (2), an almost continuous layer of green treetops. Smaller trees and a shrub layer (3) form the understorey below this. Beneath them are tree seedlings and, on the jungle floor, a thin layer of dead and rotting vegetation (4). Rivers cut through the rain forest, with dense vegetation along their banks (5).

The rain forest has several distinct layers of vegetation.

22

What animals live in a rain forest?

Every major group of animals, except those that are seawater creatures, has species that inhabit jungles. These hot, humid tropical forests contain an astounding number of animal types. Of the 8,600 species of birds that live in the world today, for example, one-fifth of them dwell in the rain forests around the Amazon river.

The animals of the rain forest tend to inhabit a particular forest layer, but are most numerous in the canopy. They live in the air above the trees, in and among the trees themselves, on the ground and in the many rivers, lakes, and swamps of these high-rainfall forests. Because of this wide range of niche possibilities, rain forest mammals, birds, fish, reptiles, amphibians, and invertebrates abound. Many species are adapted for life in trees, such as flying frogs and the bats called flying foxes.

The variety is made greater by the ways in which the lives of animals and plants become linked. Flowers may depend on birds, bats, or insects to carry pollen from one to another. Many fruits have attractive colors or smells which draw animals to eat them and so spread their seeds.

1 Brilliantly colored parrots and macaws are common birds in the topmost parts of a rain forest. They use their strong claws and beak as three "hands" for agile climbing in the branches. They feed on fruits and nuts which they gather in the canopy or sometimes on the ground.

2 There are seven species of sloths, all of which are found in the jungles of South and Central America. They have only two or three strongly clawed toes on each foot and they use these for upside-down walking under tree branches.

3 The vine snake of West African rain forests lives and hunts among the small trees beneath the main canopy. It hangs motionless among the twigs there, waiting to catch tree frogs, lizards, or large insects that move past it. Its bent shape and brown tones camouflage it.

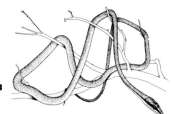

4 On the forest floor, foragers burrow for food. This giant forest hog is the largest wild pig. It uses its flattened snout to root for small animals, bulbs, and tubers in the forest leaf-litter and soil.

5 The jungles of the world have their own river-dwelling crocodiles. These highly efficient, sharp-toothed predators eat fish and mammals that enter the water.

23

What plants do you find in the jungle?

You find all types of plants, from mighty trees to almost microscopic algae that live on the surface of larger ones. There are also many fungi.

The variety of plant life in a jungle is staggering. This richness in vegetation is one of the reasons the rain forest is such an important type of ecosystem. It contains a huge treasure house of plant species that have only just begun to be tapped as sources of human food, medicines, and chemical raw materials.

The tallest jungle trees may grow up to 200 feet (70 meters) high and usually have a single, non-branching trunk. The lower layers of the forest contain many kinds of shrubs and an exotic variety of ferns. And the trees themselves act as supports for climbers, vines, and strangling figs. All these plants have roots in the ground which take up water and mineral salts.

In addition there are a number of other plants which do not have roots in the soil. These include the parasitic plants and the various hanging plants.

Ferns in La Selva rain forest in Costa Rica

Giant ginger plant in Sumatran rain forest

Bromeliad in cloud forest, southern Mexico

24

How do some plants grow without soil?

Hanging plants, or epiphytes, have no roots in the soil; instead they are anchored to the trunks or branches of trees. The whole plant, with its roots, stem, and flowers, is attached to the tree.

Epiphytes can probably live in the jungle so successfully because the air is so humid all the time. In a typical rain forest, almost all the surfaces of the larger trees will be smothered with a profuse growth of these "hangers on." The most common plants that live this apparently vulnerable type of life are mosses, lichens, bromeliads, and orchids.

Growing high above the ground, they gain enough light. Their grasping roots ensure that they are firmly attached, and they take water directly from falling rain or from water running down a tree trunk. This water can be obtained by dangling roots with absorbent outer coats (a common trick of the orchids) or stored as a pool in a cup-like rosette of leaves (a bromeliad method). The bromeliad pool, by acting as a trap for insects, can also provide the plant with minerals as the bodies of these creatures decay.

An orchid with exposed roots

25

How much soil does a jungle tree need to grow properly?

Surprisingly little. Most of the rain forests of the world grow on thin soil layers which are poor in mineral salts.

In a temperate woodland, the soil down to a depth of several feet (1–2 meters) holds plenty of mineral nutrients. This is because most such soils have only been developed since the end of the last Ice Age, about 12,000 years ago. Since they have only been weathered and rain-washed since then, these soils are still rich in minerals. Easily available water can also be reached a long way down in the soil, if the weather is dry. These two factors mean that temperate forest trees often have deep, penetrating roots.

Tropical soils are usually much older, and the huge quantity of water that has passed through them has washed out most of their nutrients. But mineral salts are available near the surface because the layer of dead leaves on the forest floor is quickly broken down by termites, earthworms, fungi, and bacteria. The surface soil is always wet due to the constant rain. Rain forest tree roots are therefore usually shallow.

On the far right is a profile of the soil and root systems of trees in a temperate woodland. The roots spread deep down to gather minerals and water.

Jungle trees (right) have shallow roots to take advantage of the surface supply of rainwater, and of the minerals present in the living plants themselves as well as in the decomposing leaf litter. The roots of the largest trees are no deeper than 18 in. (450 mm).

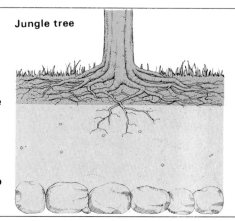

Jungle tree

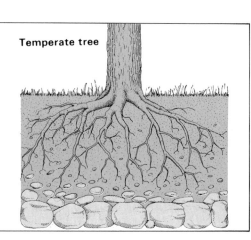

Temperate tree

26

Why are people so worried about the rain forests?

Because they are in danger of being destroyed. By 1980 about 40 percent of all the tropical forests that existed at the beginning of this century had already been cut down. In the last 30 years Central America has lost about 60 percent of its rain forests, and in Asia rain forests are being removed at the rate of 4 million acres (one and a half million hectares) per year.

These terrifying statistics show how speedily human beings are destroying rain forests. We remove them, in a process known as deforestation, because of our need for roads, land for crops, land for cattle, for mines and timber extraction, for firewood, and for hydro-electric dam lakes.

When rain forests are destroyed, it means the certain extinction of thousands upon thousands of animal and plant species. It is quite conceivable that the present rate of forest loss is condemning ten species to extinction a day, every day.

Such losses are irreplaceable. They might mean the loss to humankind of plants with enormous potential as crops, or plants which contain the next, as yet undiscovered, generation of anti-cancer drugs. If the plants go, so do the chances of ever using their products. The world would also be an uglier and poorer place without the beautiful life of the rain forests.

These forests are also a kind of protection system in areas of poor, infertile tropical soils. They regulate the flow of water to natural and human-made drainage channels. Without the vegetation cover of the forests, soil erosion and the risks of flooding increase. Probably as a result of deforestation, India now has to spend huge amounts of money on river flood defenses.

Those people around the world who fear for the survival of the rain forests and worry about the effects of their loss are concerned about a real and intensely important problem. It might already be too late to save the rain forests in some areas.

Will they regrow if they are cut down?

Only very slowly, if at all. It probably takes hundreds of years for a complex, many-layered forest to develop from a large clearing. Because the largest trees grow so slowly, it cannot be speeded up.

The forest only has a chance of regrowing if the cleared area is not too big. If a huge area is cleared, the underlying soil becomes eroded as soon as it is unprotected by the trees. This means that new trees do not have the proper soil and minerals they need to grow into forest giants.

Left undisturbed, rain forests do gradually and constantly replace themselves. Individual large trees become old, die, and fall down, often felling other smaller trees in their path. The gaps that they leave let light into the forest floor, so that seedlings of the surrounding trees start to fill the space. Rapidly growing "colonizer" trees plug the gap first. Later, though, one tree will outgrow the others and replace the tree that fell. This process is illustrated diagrammatically (right).

Timber extraction is one of the most intense pressures on the survival of rain forests. Here a logging road is being built to remove timber in the magnificient rain forests of Borneo.

A small clearing being colonized

Is the jungle noisy?

Jungles can be surprisingly noisy places. It is often raining hard in a jungle and the noise of the rain makes a constant roar. The animal life of the jungle also produces a fantastic variety of loud calls, songs, and shouts.

There is good reason for all this noisy communication. The densely growing trees of the rain forest make it difficult to see very far in any direction. If a jungle animal wants to signal to competitors, mates, or family members over long distances it can only do so with sounds.

Almost all the animal groups join in the cacophony. Birds of all types call, with loud and distinctive noises that sometimes give the birds their names. Trumpeters, for instance, live on the floor of South American jungles. High in the same forests live howler monkeys whose howls carry huge distances. Tree frogs and toads join the chorus while they sing, croak, and chirp.

The raucous calls of these brilliantly colored toucan species are part of the chorus of South American jungles.

29

Can people live in the jungle?

For rain forest tribes around the world, the plant and animal life of the forest is both home and the source of almost all their food and raw materials.

The numbers of such peoples are declining everywhere, as the tribespeople leave the forests, as they catch diseases such as measles (introduced by settlers) and as the forests themselves disappear. For instance, according to the estimates of international agencies, the Amerindian population of Brazil has dropped from about 5 million in 1500 to less than 200,000 in the late 1980s.

The pygmy tribes are scattered over much of tropical West Africa. They catch game in the forests with poison-tipped arrows shot from wooden bows, and with 300-foot (90-meter) long nets made of twisted vine fibers.

The Mbuti pygmies live in "bands," or extended families, in the rain forests of Zaire.

30

Do they catch deadly diseases there?

There are two particular sorts of disease that are caught by the peoples of the rain forests. The first are the natural, often parasitic, diseases of the forest, many of which are spread from person to person by forest insects. Examples include malaria, yellow fever, and the worm disease filariasis, which causes the swelling disfigurement of the legs called elephantiasis. All these are transmitted by mosquito bites. Eye worm, or loaiasis, is caused by the bites of blood-sucking flies.

The jungle peoples may suffer from these diseases but are rarely killed by them. They have a kind of built-in resistance.

Far more deadly to jungle tribes are the imported virus diseases brought into the rain forests by people from industrialized societies. Strains of diseases such as influenza and measles, to which jungle people have no resistance, can be deadly. In 1977 a measles epidemic killed half of the Yanomami tribe in northern Amazonia in a few weeks.

Where are the world's grasslands?

Large grass-covered areas are found in many areas of the world, both above and below the equator. Different types of grass have developed in different places but, as the map of the world shows, grasslands are an important part of the vegetation in all the large areas of land. These grasslands include: the prairies of North America, the pampas grasslands of South America, the savanna grasslands of Africa, and the steppe grasslands of Central Asia. An extreme type of grassland is the wet, barren land known as tundra, beyond the coniferous forests in the Arctic (see Question 46).

The main reason that grass grows in these places, rather than trees and shrubs, is the climate: the pattern of rainfall and temperature throughout the year. It is too dry for forests and too wet for deserts to form. There is either not enough rain each year to favor the growth of forests, or the annual amount of rainfall is quite high but it falls mainly at one time of year. This means that there are other long, dry seasons, with occasional fires, which prevent trees from establishing themselves.

Grassland areas often have good soils: deep, without any rocks and boulders in them, and with large amounts of mineral salts to provide food for plants. Because of the rich soil, and because grassy zones are usually flat, natural grasslands are often used for farming. Good, fertile soils are ideal for growing crops, and flat fields are much easier to cultivate and harvest than steep, mountainous ones.

Many of the grasslands in temperate areas are known as the "bread baskets" of the industrialized nations of the world. The grasslands of the Midwest of the United States, for example, are almost entirely used for farming cereals. Food crops such as wheat, oats, and barley are cultivated varieties of the wild grass species that existed in these grasslands, even before humans walked on our planet.

In the tropical and subtropical grassland areas, different grass types are cultivated, including millet and sorghum. Rice, the plant which perhaps feeds more people in the world than any other, is also a grass. It has the special ability to thrive in waterlogged soils.

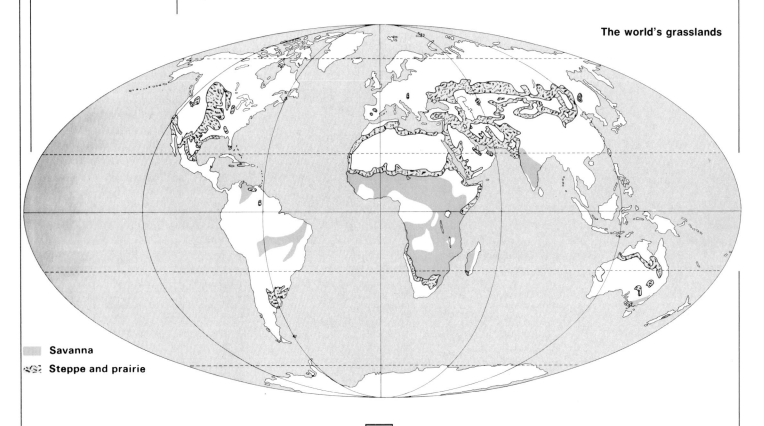

The world's grasslands

Savanna

Steppe and prairie

32

Are they all the same?

No. In all grasslands the plants that we call grasses are the dominant form of plant life, but in other respects grasslands in different parts of the planet are quite unlike one another. Even within one grassland region, there may be significant variations from area to area.

The prairies and high plains of the Midwest of the United States make up a vast grass-covered area. Before the Europeans came to this landscape, unbroken grass stretched from Pennsylvania and Ohio in the east to the foothills of the Rockies in the west. Bison and pronghorns were the large plant-eating animals (herbivores) of the terrain.

But this vast area of grassland is made up of different zones. In the east, where the soils are deep, fertile, and well-watered by rains, there are tall grasses such as the blue stem grass and Indian grass, which grow to 6 feet (2 meters) high or more. As the land rises slowly toward the west, the soil becomes thinner and poorer, while in the "rain shadow" of the Rockies the land becomes dry. Going west the typical grasses get shorter and shorter, and in the far west grow the short, hardy grasses such as buffalo grass.

By contrast, savanna grassland that predominates in tropical areas such as parts of Africa. Savanna contains isolated trees and clumps of trees and shrubs scattered among grasses such as the red oat grass. Different herbivores crop different sections of this grassland.

The prairies of southern Alberta are one type of grassland, characterized by short grasses.

33

Are grasses different from other plants?

Grasses are different from other plants in a number of ways. They have flowers, but these are neither colorful nor scented. This is because the grasses, in the open spaces of the grasslands, are all pollinated by the wind. They do not need insects and birds to carry pollen from plant to plant, therefore they do not have to attract them by means of showy flowers or powerful fragrance.

Grass flowers are clustered together in hundreds on a single grass head. They are usually small and green or light brown, and their stamens, or male parts, produce huge amounts of pollen.

Unlike most plants, a grass does not grow only from its tip, but also from a growth zone near the base of each stem section and leaf. The spreading roots, and the many new growing points near ground level, help grasses to withstand severe grazing. Even if the stalks are chopped right down to the ground by plant-eating animals, they will quickly spring up again.

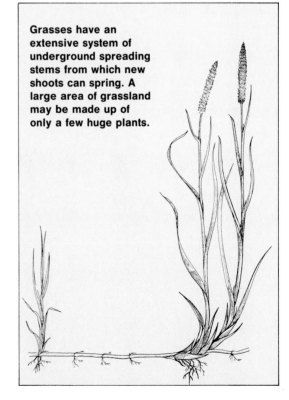

Grasses have an extensive system of underground spreading stems from which new shoots can spring. A large area of grassland may be made up of only a few huge plants.

What animals live in grasslands?

The open spaces of the world's grasslands are home to a wide variety of animals. Many of them are specialist grass-eaters, which live on the grass seeds in the heads of fruiting grass as well as the grass blades themselves.

The wide, open spaces of grassland country means that large animals can easily be seen, since there is hardly any tall vegetation in which to hide from hunters. Many of the larger herbivores, such as zebras, antelope, deer, and ostriches, are fast runners and use their speed to escape from predators.

Other grassland creatures are more skulking in their habits—mice, rats, and harvest mice live either in deep grass near the ground, or in burrows under it. The prairie dogs, a type of North American rodent, build complicated underground homes for large colonies of animals.

The easily cropped grass vegetation is also food for large numbers of insects. These include grasshoppers, locusts, and termites. Termites are social insects like ants and bees and live in colonies. They build earth "cement" nests which stick up above the level of the grass cover.

The female harvest mouse builds a ball-like nest 12 in. (30 cm) above the ground for her young. Suspended among sturdy grass stems, it is made of cut lengths of grass woven together and lined with shredded grass leaves.

The kob, a type of antelope, lives in the grasslands of Uganda in East Africa.

35

Is grass a good food for them?

Grass is not, in itself, a particularly good food for most animals. Although there are useful food substances locked up inside grass blades, they are difficult for animals to extract and use. But the herbivores which eat grass use special methods to extract nourishment from it.

Grass seeds are the most nutritious part of the plant. They contain stores of starch, protein, and fats, and animals can easily digest and use these. But grass seed is available only at a certain season, so for the rest of the year grass-eaters have to eat leaves (blades) instead.

Most grass leaves have very few proteins, starches, or fats. They are made mostly out of the tough material known as cellulose. Although this is built chemically out of the useful sugar called glucose, ordinary animals find cellulose impossible to digest; it serves as fiber, or roughage.

The second problem is the toughness of most grass blades. Grass is strengthened with silica, the hard mineral that forms flints in the ground, which makes grinding and chewing the grass leaves difficult. It is also the reason why you can cut yourself on the sharp edge of a grass blade!

36

How do they use it?

This cutaway diagram of a cow's stomach shows the inner linings of the different compartments.

Many grass-eaters have special guts for dealing with this difficult but plentiful food. Most vertebrates (backboned animals) also have teeth which are good for cropping and grinding up tough grass blades. Zebras and cowlike animals have chopping incisor teeth at the front and flat grinding molars at the back of their mouths.

To break down the tough cellulose in the grass, animals such as cows, zebras, and termites convert part of their gut into a kind of fermentation chamber. Here, bacteria help to break down the cellulose, so that some nutrients from it can be absorbed by the animal.

In the cow family this chamber is called a rumen. It is made from the bottom of the gullet. To aid the digestion of cellulose even further, cows regurgitate the contents of their rumen and chew them more finely for a second time before swallowing them again. This is "chewing the cud."

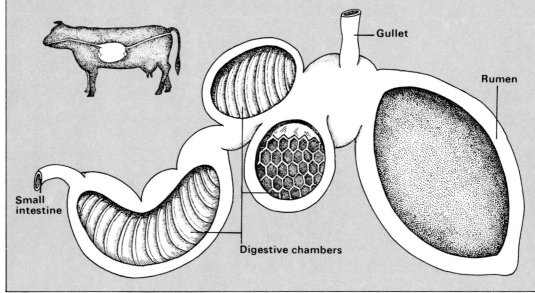

Gullet

Rumen

Small intestine

Digestive chambers

Do grassland animals eat other things?

In the African savanna large animals feed on the branches, bark, or foliage of trees as well as on grass.

Some of the grassland herbivores eat several other types of plant food besides the grass plants themselves. This is especially so in the mixed grass, shrub, and tree country of the tropics called savanna. Because there are so many different sorts of plants there, a very mixed community of plant-eating animals lives together in this landscape. It is this great variety of animals which has made the East African plains famous for their wonderful animal life.

Monkeys and giraffes can pick or bite off and eat the leaves, shoots, and buds from savanna trees. Besides its extremely long neck, the giraffe's other assets include a long muscular tongue and an upper lip that can project outward. With these working together it can grab leaves and twigs from acacia trees 20 feet (6 meters) high. With its flexible tongue it takes young leafy shoots from among the thorns and spines of the acacia branches.

Elephants also inhabit African grasslands. They will eat grass itself but can also use their long trunks like sensitive and strong hands to grasp and pull down leafy branches from trees. They will even use their massive strength to push down trees to get at upper leafy sections that are out of reach.

The Cape eland of South Africa is tall enough to feed on low branches of trees and on all bushes. It also digs with its hooves for roots, bulbs, and tubers in the ground. The shorter antelope, the gerenuk, can stand on its hind legs to reach leaves that are higher up on a tree.

The black rhinoceros, with its pointed upper lip, which can grasp at objects rather as an elephant uses its trunk, adds to its twig and leaf diet by using the tough lip to strip bark from trees and eat it. By contrast, its relative, the white rhino, has a broad mouth which it uses for grazing.

Carnivores, such as large birds of prey, are important in grassland communities because they keep down the rodents.

38

What is a dust bowl?

A dust bowl is a region of land that has been made barren, dry, and unfarmable by unthinking farmers. As wind gusts across the ground, clouds of dust swirl from the lifeless land.

In a natural grassland or a forest, the roots of all the plants help bind the top layer of soil (topsoil) and the lower, or subsoil, layer together. Held together by the roots, it is very difficult for heavy rainfall to wash away the soil.

In the same way, the parts of the plants that grow above ground act as a shield against the wind, preventing the soil from being eroded, or blown away. Trunks, stems, and leaves of plants all help to break the force of the wind.

But if these natural means of protecting the soil are destroyed, and changed by farmers into large areas covered by crop plants, there are times of the year—particularly after harvest when the plants are cut down—when the soil loses its protection. Removing a crop from a field also takes mineral salts away from the land, depriving it of nutrients. These two dangerous changes—the increased risk of the top layer of soil being eroded, and the steady loss of soil nutrients—can cause a dust bowl, in which nothing grows.

Careless farming methods produced a dramatic dust bowl in some parts of the Midwest of the United States in the first half of this century. This picture, taken in 1938, shows an abandoned farmhouse in Texas.

39

Can it be made fertile again?

Yes. If we can find out what caused the dust bowl, it is usually possible for land to be made useful and fertile again. What has to be done is to change the farming methods and to introduce ways of protecting the land, so that any nutrients lost from the soil are replaced, and the soil is not eroded. These changes cost a lot of money and need to be well planned.

Often the first stage in reclaiming a dust bowl is to stabilize the soil, that is, to prevent it being blown away by the wind or washed away by the rain. Windbreaks in the form of fences or trees can help this. Another successful tactic is to plant rapidly spreading grass species which can anchor themselves in dusty dry soils. When they are established, they help to hold the soil together.

Once the soil losses have been slowed down or stopped, the soil must be made fertile again. This can de done by adding both natural and artificial fertilizers to the ground. The farming practices themselves should then be changed to use a system of rotating crops, so that any nutrients taken out of the soil by one crop are replaced by the next crop to be planted on that piece of land. There are also ways of tilling the soil and harvesting that prevent soil erosion.

Why don't trees grow on the tops of mountains?

The peaks of the highest mountains never have trees growing on them. Part way up such mountains is the "tree line," above which no trees will survive. The actual altitude at which this "line" can be seen varies in different parts of the world. It is affected by the climate—the amount of rainfall and the temperature levels—in different places.

The reason for this has to do with the way trees grow. The growing points of trees are a long way above the ground, at the tips of branches and twigs. In really cold, exposed places, such as high mountain tops, this is a dangerous place for a plant's growth zones, since they are exposed to the worst of the low air temperatures. Tall trees, in particular, are ill-suited to life on the higher mountain slopes because their delicate growth zones cannot cope with the cold. In addition, they would not be able to withstand the heavy weight of fallen snow on their branches.

On really high mountains, the upper slopes have no plants living on them at all. This is because they are permanently covered with snow and the snow lies directly on unweathered rock, with little or no soil on it. Such conditions make ordinary plant life impossible, since all plants need water in liquid form to live and to grow. Although water exists in these places, it is present only as ice and snow. These solid forms of water are of no use to a plant because only water in liquid form can move through plant tissues.

In some parts of the world trees do, in fact, grow on mountains, as long as they are not too high and it is not too cold. Low mountain ranges in tropical regions of the world, for instance, can be covered in luxuriant rain forest.

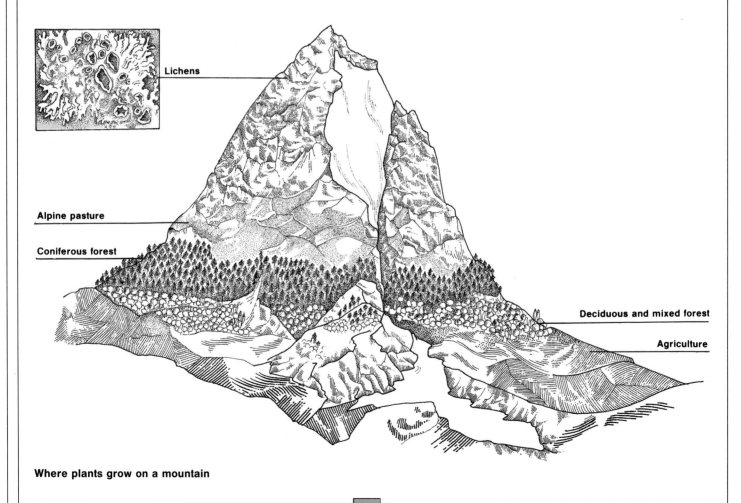

Lichens

Alpine pasture

Coniferous forest

Deciduous and mixed forest

Agriculture

Where plants grow on a mountain

Can other plants grow there?

The sorts of plants that grow just below the permanent icy wastes on the tops of mountains are tough ones. They can stand the extreme cold and the shortages of water and nutrients. Certain plants manage to get a foothold even on surfaces of pure rock, and grow there.

Among the plants that are able to live at high altitudes are the lichens. Some lichens form thin crusty sheets over the rock surface. They absorb what little water there is from falling rain or melting snow, and minerals, plus some more water, from the rock. They can also survive long periods of drying out. In these harsh conditions, they grow very slowly and live for a long time. Some mountain lichens grow only $\frac{1}{32}$ inch (a millimeter) in a year, and may live for a thousand years.

Mosses, low grasses, and heathery plants also form a layer of tough plant life. At high altitudes on some tropical mountains, such as Mount Kilimanjaro in Kenya, strange sorts of plants have evolved, forming structures like pillars. Many of them have developed from ordinary small plants like lobelia and groundsel. These adapted forms often end in a thick tuft or rosette of thick hairy leaves which protects their growing points.

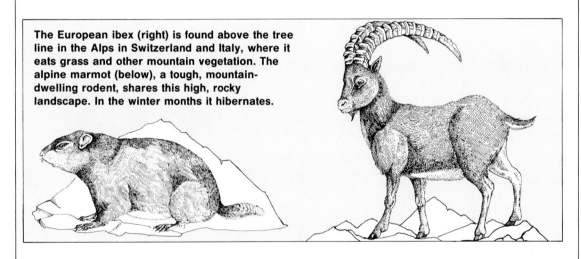

The European ibex (right) is found above the tree line in the Alps in Switzerland and Italy, where it eats grass and other mountain vegetation. The alpine marmot (below), a tough, mountain-dwelling rodent, shares this high, rocky landscape. In the winter months it hibernates.

How high up a mountain can animals live?

As high as there is food for them to eat. In the mountains, high plateaus and steppe country of Tibet, for instance, there are some large herbivores which can stand the cold and the thin air of the Himalayas. There they feed on tough, low plants.

The two animals which live highest up in this mountain country are the yak and the less well known Chiru antelope. The yak is found at heights of 14,000–20,000 feet (4,000–6,000 meters) and is well protected from the cold by its immensely thick fur coat.

On the slopes of some high African mountains, the mammals that get up highest are groove-toothed rats and rock hyraxes, which are strange, tiny relatives of the elephant. On the rock hyrax's feet are flattened nails which look like hooves. Each of these acts like a small, nonslip rubber pad to enable the creatures to run easily up the steep rock faces.

The large, soaring birds of prey also take advantage of the high mountains, feeding on both the animal life and the rotting flesh of dead creatures found there. Verreaux's eagle is a typical example from the mountains of Africa.

In the mountain peaks of the European Alps are high-country mammals and birds found in few other places in the world. The alpine chough, for instance, is an upland relative of the common chough, found on seashores and lowlands. While the lowland species has red legs and beak, the alpine form has a yellow beak.

Typical mammals of the Alps include the ibex, a relative of the mountain goat with magnificent ridged horns. It grazes on alpine meadows in the summer and, in the harsh conditions of winter, may descend to the upper edges of the forest.

Do penguins only live in cold climates?

Most penguins live and feed in the icy waters of the Antarctic region, but there are a few types that have spread their range farther north.

Penguins are primitive birds which cannot fly. They are adapted for life in the sea and use their flipperlike wings to propel them underwater while chasing their fish food. They nest on land and hatch a single egg at a time, incubating it in a warm fold of skin above their feet. Many types actually breed on the freezing ice shelf of Antarctica itself, others on nearby islands. The larger species, such as the Emperor penguin, manage to incubate their eggs in the middle of the Antarctic winter.

One species, the Humboldt penguin, has escaped the harsh climate tolerated by its near-relatives and has ended up

near the equator. It lives and breeds on the western coast of South America and feeds in the rich, fish-filled waters off the coast. The productivity of these waters is created by the cold Humboldt current. This current brings nutrient-laden waters to an area where the warm tropical seas contain very few fish and other animals.

The Humboldt penguin is named for the cold current off the coasts of Peru and Chile in which it lives.

44

What other animals live in cold climates?

Animals such as seals, polar bears, and musk oxen, as well as penguins, are all able to cope with the special problems of life at low temperatures.

Life in a cold climate presents two main problems. First of all there is the direct effect of the cold itself. Low temperatures can damage living tissue if a creature's body water starts to freeze. For a warm-blooded animal like a bird or a mammal, icy temperatures also mean that more energy is needed to keep up its body temperature. The bodies of animals that live in the coldest parts of the world are well insulated with either thick layers of fat or a thick coat of fur or feathers, which stops body heat from escaping too fast.

The second problem is finding food in the snow and ice. These freezing conditions stop most plant growth. They also make it difficult for animals to find prey or to dig food out of soil which has become hardened by ice-formation. Penguins, seals, and, for some of the year, polar bears, get much of their food from parts of the sea that stay unfrozen. Reindeer dig in the snow using their antlers, to find the plants on which they feed. Small animals, such as lemmings, survive by tunneling under the snow to find seeds and plants that are buried.

Many of the animals that live in snowy landscapes have white camouflaged bodies to help them merge into the background and prevent their being easily detected by their prey.

45

Do the same animals live in the Arctic and the Antarctic?

No, the animals that inhabit the snow-covered lands at the top of the globe are quite different from those that live at the opposite end of the world. The landscapes themselves are quite different too. The Arctic is a huge ice-covered ocean, called the Arctic Ocean. Land masses such as Greenland stretch above the Arctic Circle but most of what looks like land surface is really thick slabs of floating ice. Antarctica, by contrast, is a huge land continent, with mountain ranges, all covered in a thick ice sheet.

The common animals of the Arctic are, in the extreme north, polar bears and species of seals particular to the Northern Hemisphere. Farther south, on the fringes of the Arctic, live ermine, snowy owls, musk oxen, reindeer, and their close relatives, the caribou. None of these animals occurs in Antarctica.

In the Southern snows live penguins and four kinds of Antarctic seals. Among the best known of the Southern Hemisphere seals are the fast-swimming leopard seals. During the Antarctic summer, swimming penguins are one of the main prey animals of the speedier leopard seals.

Snowy owl

Ermine

Polar bear

46

What is tundra?

Tundra is an ecosystem that is found on land where the average temperature is very low. Even in the warmest summer months, there is still a permanently frozen layer of soil beneath the unfrozen soil surface. This perpetual layer of underground icy soil has a technical name—the permafrost.

Winter temperatures in the tundra zone can be well below freezing, and even in the summer months the air temperature is never very high. The rise in temperature, though, is enough to thaw the top layer of the previously frozen ground, and this thin layer of thawed soil, with water in it, supports the growth of some tundra plants.

Only very tough, cold-tolerant plants can make growth and produce flowers in the short summer season. Solely because their growing points, containing buds, are so close to the ground can they survive

through the cold, harsh winter months, sheltered from the worst of the wind-driven snow and ice particles.

Tundra landscapes are almost completely restricted to the northern half of the world. They are found mainly in a band based roughly on the position of the Arctic Circle and north of the tree line, stretching across northern Alaska, northern Canada, around the coasts of Greenland, and across the top of Scandinavia and the Arctic regions of the USSR. North of this Arctic tundra is the ice and snow of the Arctic cap itself.

Alpine tundra, on the other hand, can form near the top of any mountain which is sufficiently high to experience cold enough temperatures to produce permafrost. This is found in the Alps in Europe, the Rocky Mountains in North America, and even on tropical mountains!

47

Does anything live there?

Besides their characteristic low, tough vegetation, tundra lands are also the home of some equally resilient animal species. They have to find ways of surviving through winters which last for eight to nine months of extreme cold.

Their methods for doing this involve good insulation against the cold, whether

Reindeer use their antlers to dig through the snow in order to find low-growing food plants, especially lichens.

in the form of thick fur or feathers, and thick reserves of body fat, built up during the summer months when food is plentiful. Musk oxen and brown bears are well adapted for living in tundra landscapes, with their thick fur coats to keep them warm and dry.

Other typical large herbivores of the northern tundra are the caribou and the reindeer. Caribou are wild deer which inhabit the North American tundra in the summer months, and move south to the forests in winter. Reindeer are their old world counterparts.

Whereas large animals such as musk oxen have to stand the low air temperatures above ground, smaller animals can, to some extent, escape these extremes by living below the winter snow layer and in, or under, the vegetation mat. The mice, voles, and lemmings of the tundra often spend the winter months in this way, by tunneling under the snow. Hidden from predators, they eat their way through the seeds and other plant matter produced during the previous growing season.

A large proportion of typical tundra species are migrating birds, such as geese, that fly elsewhere to escape the harshest winter months.

Are there plants in the Arctic and Antarctic?

Plants do live in these areas but there are not many different sorts.

The difficult growing conditions mean that in mainland Antarctica there are only one or two kinds of flowering plants. These are a type of tough grass and another sort of plant that forms wind-resistant, low cushions close to the rocky soil. All the other land plants of Antarctica are simple plants like mosses, liverworts, and lichens which do not have flowers.

The tundra landscapes around the fringes of the Arctic grow quickly in the brief Arctic summer. Here, there is a wider selection of plant types. Arctic plants include heathers and some dwarf trees only a few inches high, as well as lichens, mosses and sedges. The tall lichen called reindeer "moss" is an important food for the animals for which it is named.

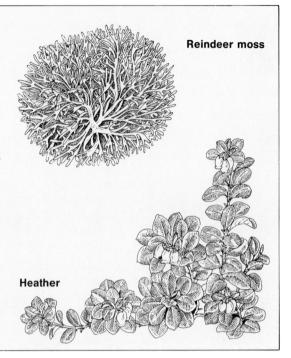

Reindeer moss

Heather

The Eskimo tribes are adapted to life in the Arctic.

Do people live in the Antarctic and the Arctic?

The only people who live in the Antarctic are scientists from different countries who keep permanent research stations on the Antarctic ice cap. They can only survive there with food supplies and fuel brought in from outside.

The story of people in the Arctic is rather different. The Inuit (Eskimo) tribes have produced a specialized way of life, using the tundra and the seas beside it for their survival. They now live in the most northerly parts of Alaska and Canada, as well as in Greenland.

Although most Eskimos now live town lives, in the recent past they were all expert "hunter-gatherers," able to live off the animal and plant life of these difficult cold latitudes. They killed whales, seals, and walrus at sea, even through the ice, using spears, harpoons, and bows. Musk oxen and caribou provided food on land. Their winter houses, known as igloos, were made out of ice, which, strange as it may seem, is a good insulating material.

50

Where are the world's deserts?

Deserts cover more than a third of the Earth's surface. They are found in North and South America, in Africa, in Central Asia, and in Australia. Deserts are the driest places on our planet and usually have less than 10 inches (250 millimeters) of rain a year. Most deserts are hot all the time, too, but not always. The Gobi desert in Asia, for example, has very cold winters, with snow and freezing winds.

What a desert looks like depends on the type of its soil, how high up it is, the climate, and the kind of plants that grow there. Where it is depends on the weather pattern. Rain-shadow deserts, for example, are situated on the sheltered side of mountain ranges, where the mountains act as a barrier to rain. Examples include the Great Basin in North America, the Sahara in Africa, and the Gobi in Asia. Most of the Sahara is in fact dry mountain and rock gravel country.

The cluster of desert lands called the Great Basin is the biggest desert in the United States. It is high-country desert, most of it about 4,000 feet (1,200 meters) above sea level, and it stretches between the Rocky Mountains in the east and the Sierras in the west. These two mountain ranges running north to south stop the winds from carrying rain into the desert country. The rain falls on the rising land on each side of the desert, leaving the middle dry.

In the southwest of the United States, the Great Basin connects with lower-altitude dry regions: the Mojave desert and Death Valley. Death Valley is famous for its fierce and dangerous climate, and gets its name from the many early European settlers who perished trying to get through it. Air temperatures as high as 134° Fahrenheit (56° Celsius) have been recorded there.

Other deserts are dry because they are close to the coast, where cool ocean currents make the winds drop their rain before they reach the land. This is the situation in South America, where a long stretch of the western coastline fronts a strip of desert called the Atacama, between the ocean currents of the Pacific and the huge mountain wall of the Andes. The Namibian and Kalahari deserts in Southwest Africa, and the desert in the west of Australia, are also created in a similar way—the cooling effects of the ocean current.

The deserts of the world

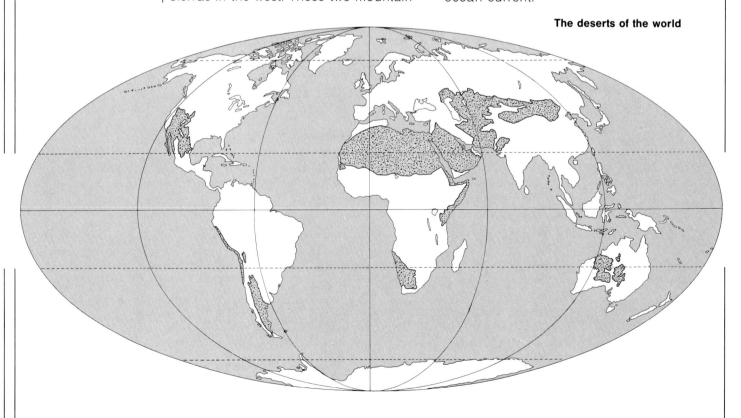

51

How do plants grow in the desert?

Desert plants have a whole number of special "tricks" which help them survive in a very dry climate. To start with, most of them have only small, spiny leaves. These prevent too much water evaporating from the leaves. The compact shapes, without large, flat surfaces, of desert cacti also helps cut down water loss from the plants.

Most desert plants can store large amounts of water, either above ground in their stems, or below ground in roots or tubers. These "bags" of water can be used between the rare times of actual rainfall.

Many desert shrubs have deep and wide-spreading root systems to catch every drop of precious water in the ground. Individual plants also grow spaced widely apart, which means that they do not need to compete with each other for the small amount of water there is in the soil.

The giant cacti of the North American deserts may reach 50 feet (15 meters) in height and weigh as much as 10 tons (100 kilograms). Many desert bird species nest in them, just as birds nest in trees in temperate countries.

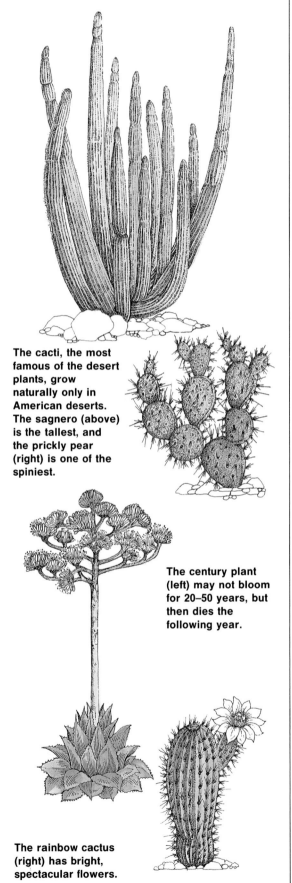

The cacti, the most famous of the desert plants, grow naturally only in American deserts. The sagnero (above) is the tallest, and the prickly pear (right) is one of the spiniest.

The century plant (left) may not bloom for 20–50 years, but then dies the following year.

The rainbow cactus (right) has bright, spectacular flowers.

52

Do desert plants ever flower?

Yes they do. Almost all the plants that live in deserts are in fact flowering—their seeds are produced inside flowers. And all flowering plants—even desert ones—bloom at some time in their lives.

Most of the time, it is true, there are no flowers to be seen in a desert. Unlike forests or grasslands, where there are some plants in flower all year round, deserts only have the right conditions for making flowers for a short time in the year. Sometimes the gap between flowerings is more than a year long.

In deserts such as the Arizona desert, where there is a small amount of rain once a year, most of the flowering occurs just after rain has fallen. At that time plants have enough water to make flowers, fruits, and seeds. And it is probably also when insects or other animals are available for flower pollination.

In very dry deserts, rains may not fall for several years. But when the rains come, the desert blooms in a matter of days.

Can animals live in deserts?

Certainly. Wherever plants can grow, you will always find animal life as well. The deserts are no exception to this rule. The plants of the desert are eaten by desert herbivores and they are, in turn, eaten by several types of desert carnivore.

Although the desert has the full range of the main animal groups in it—including mammals, birds, and insects—the only desert survivors are those creatures that can live and breed in the harsh, dry, and often hot conditions of the desert.

A camel, one of the largest animals to inhabit the desert, shows many of the special adaptations a creature needs to cope successfully with life there. It has wide, flat feet to stop it sinking into soft sand. Its hump is a foodstore of fat to help it to survive long dry seasons. When a camel gets the chance, perhaps at an oasis water hole, it can drink a huge quantity of water—almost 15 gallons (70 liters) at one go. It conserves body water carefully too; even camel droppings are ultra-dry, to help the animal conserve precious water. Other desert creatures have their own special tricks for desert life.

The western spadefoot toad lives in North American deserts and uses its spade-shaped rear feet for digging. For 11 months of the year these toads may live underground, encased in dry mud. They come out to feed and mate in the temporary pools formed after occasional rainfall.

The fennec fox is found in the deserts of northern Africa. It hides in cool sand burrows in the heat of the day. Its huge ears may act as radiators to get rid of excess heat. They are also sensitive sound antennae for picking up the tiny underground noises of the insects on which these foxes live.

The northern three-toed jerboa inhabits the deserts of central Asia. It spends the daytime in a burrow and feeds on seeds and insects at night.

The sidewinder snake of North American deserts has a special sideways means of locomotion, which is ideal for getting over hot, shifting sand. Its body has to touch the sand at only two points at a time for this movement to work. The sidewinder snake feeds at night on lizards.

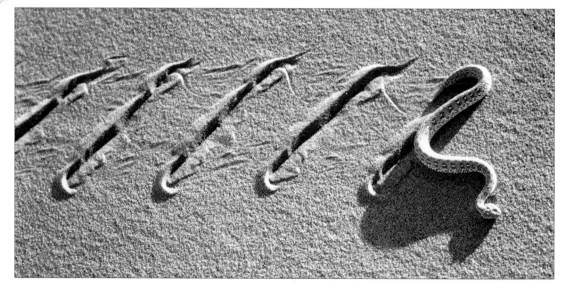

54

What is an oasis?

Even in the dry wastes of desert country there are pockets of green vegetation and standing water, known as oases. Plants grow luxuriantly around them, in sharp contrast with the barren desert beyond.

Oasis water hardly ever comes from local rainfall, since most of the rain that falls in the desert evaporates immediately in the heat of the day. Instead, these oasis water holes and pools are fed from underground or distant water sources.

Some oases are the result of natural cracks or faults in the top layers of rock which allow water in the water-carrying (permeable) rocks below to come to the surface. In others the ups and downs in the layers of permeable rocks bring them to the surface here and there. At these places water gathers and an oasis forms.

Wherever these springs exist—and the best known examples are in the Sahara desert in North Africa—oases become amazing concentrations of plant and animal life. Date and other palms cluster thickly around the water's edge. Animals such as camels make long journeys to the oases to collect water and food. Sand grouse may make round-trips of more than 60 miles (100 kilometers) to get water for their nestlings at an oasis water hole.

Artesian wells are bore holes sunk down by humans to the water-carrying rocks. These wells can be used in deserts to produce artificial oases. Farmers have also recently expanded some existing oases, by drawing water up from underground, in order to irrigate the surrounding land.

The diagram shows how the water found in an oasis is really spring water. It collects in the pools which form when water-carrying (permeable) rock layers come near to the land surface and are surrounded by water-blocking (impermeable) rocks. The permeable rock acts like a hard sponge, and water can travel huge distances sideways through rock of this sort. This movement means that water can flow underground from areas (perhaps by the sea) where there is more rainfall, to the site of the oasis.

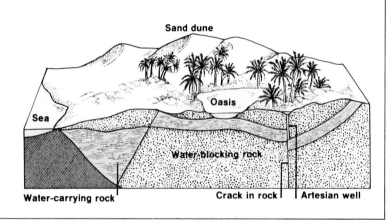

Sand dune

Oasis

Sea

Water-blocking rock

Water-carrying rock

Crack in rock   Artesian well

55

Are deserts really growing?

Many of them are. Because of slow changes in the climate pattern of the Earth, deserts in different continents may expand or shrink. In North Africa at the moment climate changes are making the edges of the desert spread into nearby non-desert lands. This change is called desertification.

Desertification means that any sort of farming becomes more and more difficult, and the spread of the desert in this way eventually brings starvation. It is happening at an alarming speed on the southern edge of the Sahara desert in the Sahel zone. The rains failed for several years running and the savanna grasslands of the Sahel region could no

longer provide enough food for cattle. Much of the Sahel savanna is now desert—an extension of the Sahara. This has produced enormous problems for the people who live in this area of Africa, and thousands have died of starvation.

As well as natural climate changes, over which we have no control, it seems likely that human actions may sometimes speed up the expansion of a desert. Overgrazing grassland or savanna with too many domesticated animals, such as cows and goats, may help the desert to spread by removing the soil-protecting vegetation. Cutting down too many trees to use for fuel or timber may have the same result once rain washes away the soil.

What are wetlands and how do animals live there?

Wetlands are the places in the world where open water and land meet. This may be around the edges of lakes, along the sides of rivers, on the banks of estuaries where rivers flow into seas, or along seashores themselves. In these places, waterlogged soils and much water make habitats called marshes and swamps. They are both types of wetland.

Flamingos feed in the shallow waters of marshes or around lake edges. They eat microscopic animals and plants that live in muddy water and soft mud. By placing their beaks upside down in the water, they can filter out these tiny food particles from the water, using the fine plates and "frills" on the inside of their beaks and on their large tongues.

In different ways, wetlands are like both land and water landscapes. Many sorts of animals and plants live there. The water itself provides a home for large numbers of fish and invertebrates, the mud or soft soil is packed with burrowing worms, shellfish, and other animals, while larger mammals, birds, reptiles, and amphibians move among the vegetation.

The animals of the wetlands often have special ways of finding food in their marshy surroundings. The unusual beak shape of flamingos, for instance, is linked with the way they get their meals. Storks, herons, bitterns, and ibises are all birds that stalk among the reeds and rushes of swamps around the world. They hunt for their prey, and catch fish, small mammals, and frogs as food. They all have long, flexible necks and pointed beaks for this hunting.

Lily-trotters (jacanas), in contrast, catch insects and other small prey on the top of floating water plants. They can walk safely over the leaf pads without sinking, using feet with four long toes which spread their weight well. Shoebills, found in African swamps, use their extraordinary large beaks to pull out lungfish which lie coiled up in protective chambers under the wetland mud.

The wealth of wildlife in the Okefenokee swamp in Georgia includes several types of heron.

How do plants survive in the wetlands?

A mangrove tree showing exposed buttress and aerial roots.

Each type of wetland has its own particular mix of vegetation, all of it adapted to the marshy habitat in which it lives and grows. An inland bog in moorland in a temperate country will have mosses and rushes, whereas reeds, papyrus, and water lilies will grow in an African tropical swamp in rain forest country.

One of the most extraordinary sorts of wetland vegetation is the tangled growth of trees and shrubs of a mangrove swamp. Mangrove swamps are found at the ocean's edge in tropical and subtropical lands such as Southeast Asia. The trees can not only survive the saltiness of the water and the effects of the tides, but they have an elaborate system of roots which allows them to anchor themselves in the soft mud— usually river silt deposited along the shore by tides and currents. These roots are shaped like great props or buttresses, and are immensely strong.

The mud of the swamp contains very little oxygen, a gas the plant roots need to keep on growing. But some sorts of mangrove trees have conical "aerial" roots that grow upward out of the water to take in air.

Can the same animals survive in both freshwater and seawater?

Most water animals spend their entire lives either in freshwater or in the sea, but some can live in both types of water. Certain fish are born in the sea but spend part of their lives in freshwater rivers and lakes inland. Others do the reverse: they are born in freshwater but later migrate to the sea. To cope with these alterations in their surroundings, the fish have to be able to adapt to great changes in the saltiness of the water around them.

European eels, for example, breed in the waters of the Sargasso Sea on the west side of the North Atlantic. The eggs hatch into tiny larvae which migrate across the ocean to reach the mouths of the rivers of Europe. This takes about three years, by which time the "elvers" are 3 inches (7.5 millimeters) long. They then move into freshwater and spend the next five years or more feeding and growing in rivers and lakes. At the end of this time, when they are completely adult, they return to the sea and swim back across the Atlantic before they die.

Young salmon, on the other hand, migrate from the freshwater rivers where they were spawned and spend their first few years swimming to the open sea. After two or three years feeding and growing in the sea, they make the return journey to the waters in which they were born, in order to mate and lay eggs. Most then die, exhausted by a journey of up to 2,000 miles (3,500 kilometers) without feeding.

How the salmon navigate precisely enough to return within a few feet of the site where they were spawned is one of the mysteries of nature. But we know that they do this by tagging young salmon and plotting their outward and return journeys.

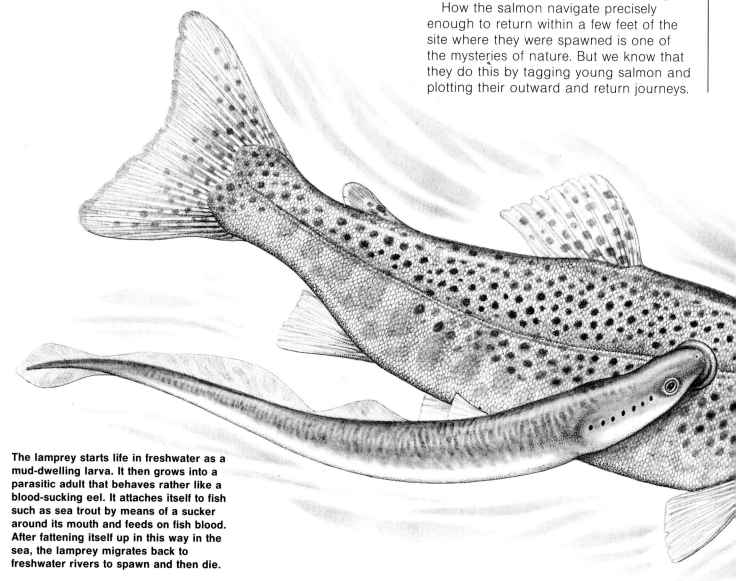

The lamprey starts life in freshwater as a mud-dwelling larva. It then grows into a parasitic adult that behaves rather like a blood-sucking eel. It attaches itself to fish such as sea trout by means of a sucker around its mouth and feeds on fish blood. After fattening itself up in this way in the sea, the lamprey migrates back to freshwater rivers to spawn and then die.

What sorts of animals and plants live on the seashore?

Almost every animal group has members that live successfully on shores. Otters hunt for fish there. Birds such as gulls and waders search for shellfish, worms, or the dead bodies of small animals. The rocks are home for a wide range of creatures which anchor firmly on them to avoid being washed away by the waves. These include sea anemones and mussels.

The reason that seashores usually teem with living things is the plentiful supply of sunlight, water, and mineral salts from the sea. These enable the plants there to grow well, and they in turn provide food for many animals—some in the water, others burrowing in sand, or attached to rocks.

Seashores are either rocky or made of shelving mud or sand. Different species of animals live on the two types of shore. The animal life is not so easy to see on sandy shores since many of the creatures live below the sand, not on it. Rocky shores have more lush plant life, most of it seaweeds. In addition, lichens can be found at the top end of rocky shores, dune plants at the top of sandy ones.

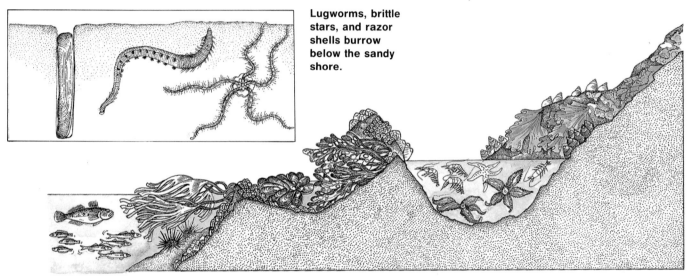

Lugworms, brittle stars, and razor shells burrow below the sandy shore.

On the lower shore are larger barnacles, sea urchins, and kelp.   **On the middle shore are limpets, mussels, and bladderwrack seaweed.**   **In the rockpools are shrimps and starfish.**   **On the upper shore are lichens and small snails.**

How do seashore animals stand being wet then dry?

They all have tricks of behavior which mean that they only really expose themselves when conditions are wet.

On any normal seashore the tide comes in and goes out again about twice in every 24 hours. So the animals attached to the rocks are in the air twice and submerged twice every day. But almost all of them can protect themselves from drying out at low tide.

Barnacles, for instance, open their hard protective plates when the tide is in and feed with feathery filtering limbs. As the tide drops, they stop feeding and close the plates so that no water is lost. Most sea anemones withdraw their tentacles when the tide goes out.

The limpet lives on rocky shores. At low tide it clamps itself to a hollow in the rock using a powerful muscular foot. As it is covered by the incoming tide it crawls over the rock surface, feeding as it goes by rasping away at microscopic plants on the rock with its rough tongue. Just before it is uncovered again, it reattaches itself to its home spot.

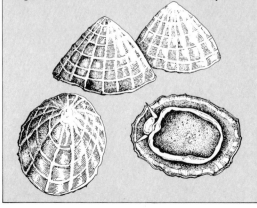

61

What is the difference between river fish and sea fish?

The answer to this question lies in the way the two types of fish deal with different conditions. The saltiness of the sea is the most important difference.

The codfish group, including important foodfish like cod and burbot, is totally marine—every one of its 700 species spread around the world lives in the sea. Most of the "cartilaginous" group of fish—sharks, dogfish, and rays—is also restricted to the sea. These fish can only live in saltwater. Their bodies are "locked in" to a life in salty conditions.

Sea-dwelling fish must be sure that they are protected from the damage that would result if too much salt were to seep into their bodies. The kidneys of sea fish are especially designed to get rid of salt from the blood. In addition, many sea fish have a gland at the end of their gut which helps keep their bodies free of excess salt.

Other fish groups have an opposite preference for water conditions. Freshwater fish are not adapted in this way and would die if they found themselves in the sea.

The flesh-eating piranhas of South America belong to the freshwater group of fish known as characins.

62

What is an estuary?

An estuary is the wide lower end of a river where it flows into the sea. This part of the river is tidal, so the water level goes up and down with the sea's tides. Estuaries are also places where there is a changing pattern of saltiness in the water. In the middle region of an estuary the water may be almost completely fresh at low tide but virtually all seawater at high tide. In some parts of the estuary there is brackish water—that is a mixture of salt and freshwater.

These continually changing conditions mean that a special mixture of animals and plants is found in estuaries. It is different from that found in the sea, and from that in the upper, freshwater parts of a river. Some estuary animals are very tolerant of changing salt levels. The flounder, for instance, is a typical European estuary-living fish. This flatfish lies on the river bottom and for most of its early life can live happily in almost any mixture of seawater and river water.

In England's River Thames there is a freshwater shrimp (amphipod) found in the upper regions of the estuary and another completely marine shrimp at the sea edge. A third form, which is particularly good at surviving in middling salt concentrations, lives in the region of the estuary between the other two.

63

Why are some seaweeds green and others red or brown?

Seaweeds, like most other plants, all contain the green substance called chlorophyll. But in seaweeds that live in deep water, the green is masked or shaded by other colored substances which may be brown, orange, or red.

If you walk from the top to the bottom of a rocky shore at very low (spring) tide, you will see a wide variety of seaweed colors. At the top of the beach, especially where there is freshwater trickling down, there is likely to be a carpet of bright green seaweeds; in these the chlorophyll is easily seen. On the middle and lower parts of the shore are many brownish seaweeds. And scattered in the rockpools and at the lowest part of the shore you will find reddish seaweeds.

Scientists think that the color differences found in the deeper-water seaweeds help them to absorb light more efficiently underwater.

Seaweeds display a wide range of colors.

What is plankton?

Plankton is the name given to all the small animals and plants that are suspended in the waters of oceans or large lakes. These organisms do not float to the surface or sink to the bottom. Most plankton is extremely small and can only be seen properly through a microscope or magnifying glass.

Although it is so difficult to see, plankton is an absolutely vital part of life in water. It provides the food for other creatures, and in doing so forms the base of most of the food chains that are found in the sea (see Question 6).

The plant plankton, technically called phytoplankton, consists of huge numbers of minute plants, each one made of a single cell or a few cells. Whenever sunlight penetrates enough, they can use it to photosynthesize (see Questions 3 and 4). Most plant plankton is found near the coastlines, where the water is shallow, and in cold currents, where there are plenty of mineral salts and other nutrients. The phytoplankton is the most important form of life in the oceans. All the other life in the sea depends on its ability to use sunlight energy to make new living material.

Vast numbers of minute animals, called zooplankton, make up the other half of

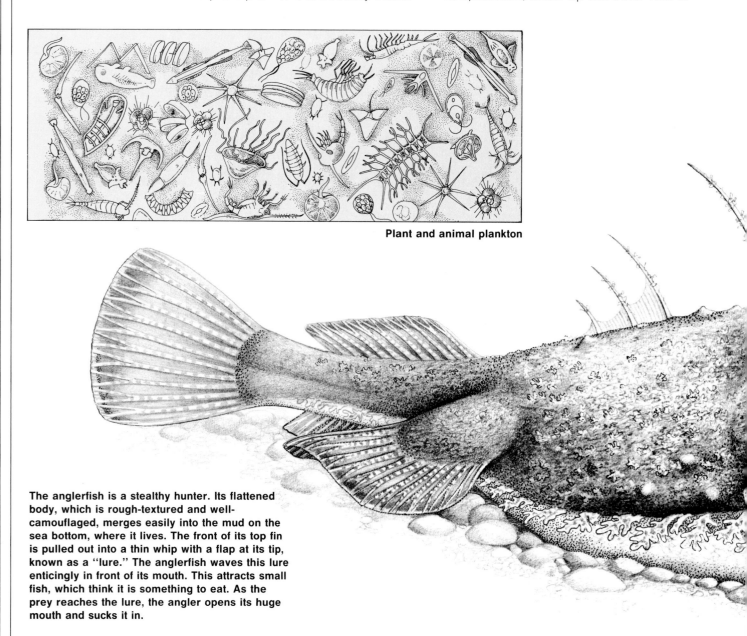

Plant and animal plankton

The anglerfish is a stealthy hunter. Its flattened body, which is rough-textured and well-camouflaged, merges easily into the mud on the sea bottom, where it lives. The front of its top fin is pulled out into a thin whip with a flap at its tip, known as a "lure." The anglerfish waves this lure enticingly in front of its mouth. This attracts small fish, which think it is something to eat. As the prey reaches the lure, the angler opens its huge mouth and sucks it in.

the plankton. The zooplankton contains: the eggs of many fish species; the very small young fish (fry) which hatch from them; many different types of small crustaceans such as shrimps; swimming larvae of marine snails; worms, starfish, and sea urchins; minute jellyfish; and small, soft-bodied hunters such as arrow worms and sea gooseberries.

These small animals eat the plant plankton or each other. They are, in turn, the prey of slightly larger carnivorous animals like fish. Some mammals, such as baleen whales, feed entirely on the small shrimplike crustaceans known as krill.

What sea creatures eat each other?

The teeming life of the seas includes an almost endless variety of carnivorous animals. These "meat eaters" range from the microsopic arrow worms to the largest animal that has ever lived on planet Earth—the blue whale.

In fact, the minute, transparent arrow worm, only a fraction of an inch (a millimeter or two) long, and the blue whale, weighing in at over 100 tons, are examples of the two major types of animal-eating sea creature. These two types are: the fast hunters of single prey animals (such as the arrow worm), and the filter-feeders which strain out small suspended food animals from the sea (such as the blue whale).

Arrow worms zoom around in the plankton to catch single small crustaceans with their sharp jaws before eating them. This is the same basic method as that used by all the other fast-swimming carnivores of the oceans, although many of them are very much bigger. Sharks chase fish before catching them in jaws packed with razor-edged teeth, seals hurtle through the sea to grasp penguins and fish underwater, while the seals themselves are chased and eaten by the even faster-swimming killer whale.

Other carnivores of the sea use more stealthy tactics—they lie in wait before pouncing on their prey. An octopus will lurk in a crack in a rock, keeping watch with its large eyes. When a crab comes into view a suckered tentacle will snake out to enfold the crab and pull it back to the beaked mouth of the octopus.

The blue whale—like the basking shark, the sea squirts, the barnacles, and most shellfish—filters its food. The filter that the whale uses is in its mouth and the food it filters is krill—shrimplike crustaceans in the animal plankton.

The whale opens it vast mouth and fills it with water plus krill. It then shuts its mouth and, with its tongue, forces the water out through a curtain of fibrous, frilly whale bone (called baleen) that hangs down around the edge of its upper jaw. This filter lets the water out but keeps back the krill which the blue whale then swallows.

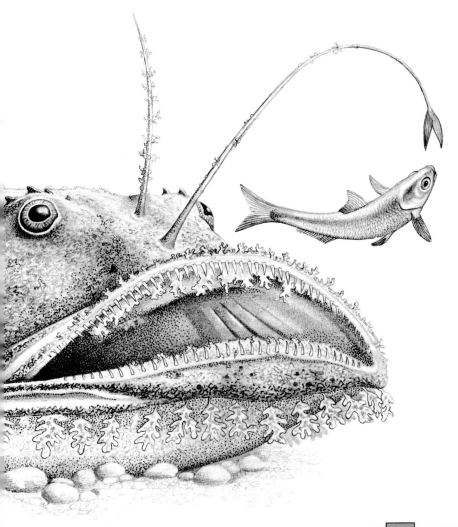

Where do you find coral reefs?

Coral reefs are found in the warm seas of the tropics. As the map shows, reefs are common around the tropical islands of the Indian and Pacific Oceans. In the same region is the famous Great Barrier Reef in Australia. There are corals down sections of the East African coast and they also stretch as far north as the northern end of the Red Sea. In the New World, there are many beautiful coral reefs in the Caribbean.

Coral reefs are massive underwater structures made of rocky material similar to that which makes up marble or chalk. But this "rock" is built by living sea creatures—the coral animals known as polyps. They are closely related to sea anemones and jellyfish.

There are three different types of reefs found in the warm waters of the world. They are called barrier reefs, fringing reefs, and atolls. The Great Barrier Reef stretches for 1,250 miles (2,000 kilometers) down the eastern coastline of Australia. It is a huge reef that runs parallel to the shoreline, but is separated from it by a considerable distance of 60 miles (100 kilometers) or so.

Fringing reefs are usually much closer to the shore. Most are found around islands in the middle of the ocean.

Atolls are circular reefs with no island in the middle. Instead, in the center of the reef ring is a shallow lagoon. Scientists think that an atoll is formed when an island with a fringing reef slowly sinks (or the ocean level rises.) The island disappears beneath the waves because it cannot grow upward. The reef, though, is living and can grow up to keep pace with the changing sea level.

Coral reefs are found only in warm saltwater and in shallow, well-lit seas. They need clean seawater, unmuddied by silt or sediment. This is why reefs are never found at really great sea depths, nor are they found on the coasts near river mouths and estuaries where there is too much silt.

There is a greater kaleidoscope of sea creatures living on a coral reef than in any other marine habitat. Apart from a host of colorful tropical fish, the animal life includes starfish, sea anemones, sponges, sea urchins, and crustaceans.

The coral reefs of the world

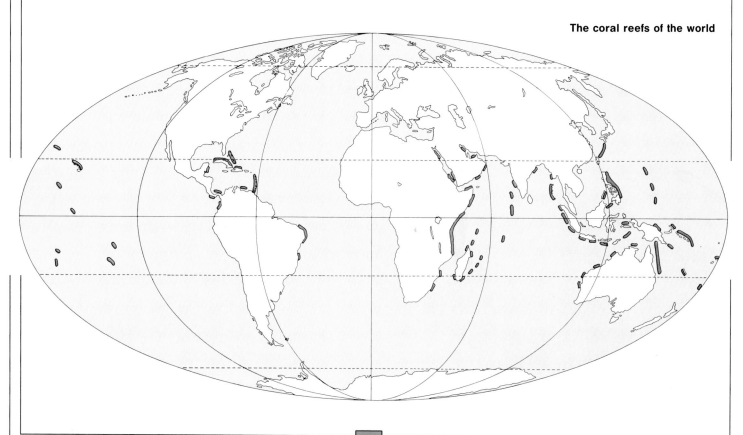

What are coral reefs made of?

Coral reefs are made from the rocky skeletons of coral animals, or polyps. Each creature inside the skeleton is rather like a minute sea anemone. The polyps live together in huge colonies of many thousands of creatures. This mass of animals, unlike ordinary sea anemones, makes a hard rocky base under itself, and around each of the polyps, as protection.

The polyps themselves poke out from tiny holes or pores in the rocky structure to feed with their tentacles. They do this by stinging their prey, then curling their tentacles over to capture it.

Vast numbers of these coral masses, or heads, fuse together to make a reef, just under the water surface. Reefs take thousands of years to form.

Most coral reefs are found in shallow water because of a strange partnership that exists in all corals. Tiny plants (algae) live inside the coral animals. Without these plants, the corals cannot make their rocky skeletons. Only in shallow waters is there enough light for the algae to live and to trap sunlight energy in photosynthesis.

Corals come in a wide variety of shapes, sizes, and colors. Some are smooth, rounded structures, such as brain corals. Others stick up into the water in complicated branching shapes, such as the stag corals resembling a deer's antlers.

Many-colored coral from a reef in the Pacific, off Papua New Guinea

The crown of thorns starfish is a serious predator of the soft-bodied animals (polyps) that construct coral reefs. Large numbers of these many-armed creatures have already caused immense damage to the Australian Great Barrier Reef. They are also found in the Red Sea, and are related to the sea urchins, brittle stars, and sea cucumbers.

This starfish, which may measure up to 16 in. (40 cm) across, can feed on the polyps despite the rocky skeletons that protect them. They do this by settling on the coral surface and turning their stomach inside out onto the polyps. In this position they pour out digestive juices from their gut. This digests the polyps and kills the coral. The digested material is then sucked back into the starfish and used as food.

Does anything live at the bottom of the sea?

There are many animals and plants living at the bottom of the sea. What they are, and what they look like, depend on the depth of the particular sea bottom. In the shallow seas of the shorelines and the "continental shelves" (the shallow waters next to the shores of many continents) some sunlight filters to the ocean bed. This enables sea-dwelling plants (algae) to trap energy by photosynthesis. These are then eaten by ocean animals, and form the base of a food chain.

Below about 3,000 feet (1,000 meters), the pattern of sea-bottom life alters. Since sunlight cannot penetrate this depth of water, the sea bed therefore has no living plants, either large or microscopic.

Instead of plant life, the foundation of most deep-bottom food chains is the constant shower of animal and plant remains that fall from the waters above. The bottom-dwelling animals depend on this continual rain of dead and dying food. Scavenging fish, brittle stars, starfish, sea urchins, lobsters, shrimps, and crabs eat this "detritus."

Other sea animals, such as deep water sharks, squids, octopus, anglerfish, gulper fish, and viper fish are deep-sea hunters that capture other animals as food.

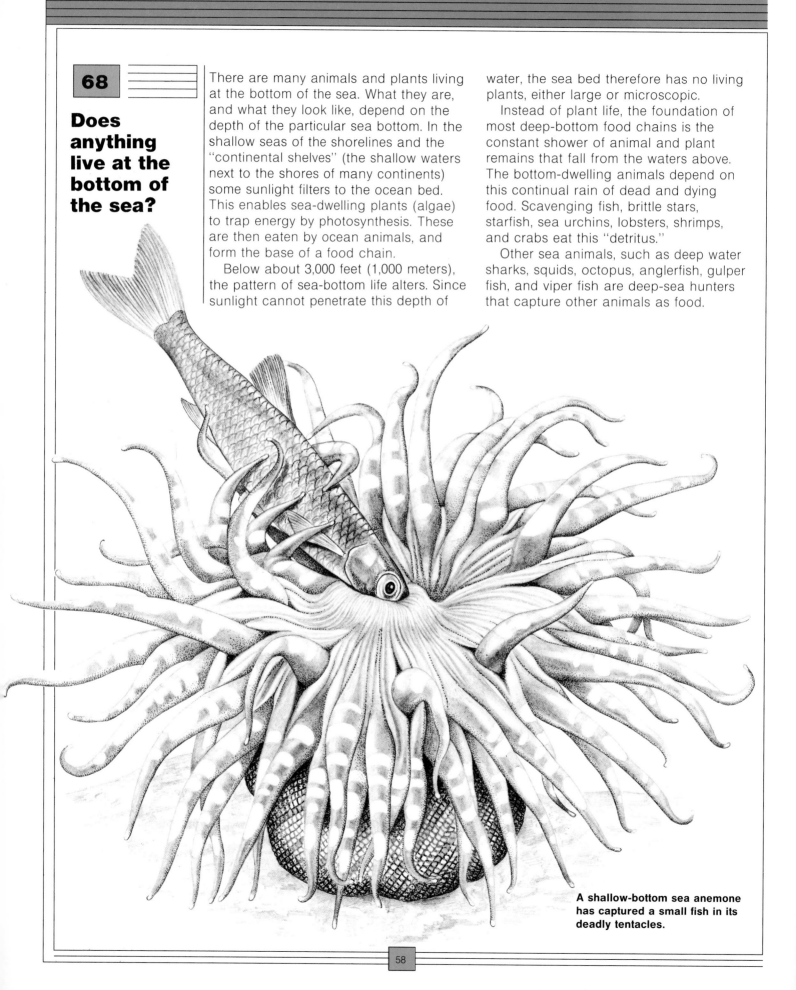

A shallow-bottom sea anemone has captured a small fish in its deadly tentacles.

Is it cold there?

It is cold at the bottom of the deeper parts of the ocean. Just as sunlight gets absorbed by the water, so too does the sun's heat. Near the sea bed, below 3,000 feet (1,000 meters), the water temperature is only just above freezing point, even in the oceans of the tropics.

The only exceptions to this are the recently discovered deep-sea thermal vent areas. These are found in the middle of oceans, near to the places where new molten sea-bed rock, rather like volcanic larva, is erupting from beneath the Earth's crust. In some spots on the sea bed, seawater spouts from the hot rocks through openings, or vents. This seawater is rich in minerals and such chemicals as the "bad eggs" gas hydrogen sulfide.

For a short distance around the vent the water is many degrees warmer than freezing. In addition, the chemicals in the warm water make possible the vigorous growth of specialized bacteria which can use sulfur, rather than sunlight, to supply them with energy. These, in turn, are the base of a food chain that includes gutless tube worms, mussels, and crabs. These animals cluster thickly around the vents themselves but are found nowhere else on the deep-bottom sea bed.

A luminous viper fish

Crustaceans and mussels in a thermal vent community

How dark is it at the bottom of the sea?

Below 3,000 feet (1,000 meters) deep, the waters of the world's oceans are totally without sunlight. This region of the deep is called the aphotic zone, which means "without light." Despite the lack of sunlight, though, the sea bottom and the deep waters themselves are not completely dark.

A wide variety of deep-sea animals responded to the challenge of pitch blackness by developing ways of making illumination. There are luminous fish, shrimps, and squids in these dark waters. Sometimes the light is produced by clusters of light-making bacteria which the larger animal keeps as partners. In others the animal can make its own "biological" light by an internal chemical reaction.

The lights on these animals serve several purposes. For some fish and crustaceans, they are signals that identify the species—the pattern of light enables mates to recognize each other. For other fish the lights aid their hunting—one anglerfish attracts its prey by means of a luminous "lure" on the front of its top fin.

Some deep-sea squids can produce luminous ink. Just like the black ink made by shallow-water octopus and squids, the shining ink of the deep-sea forms is used to confuse predators when they attack.

Why are tropical fish such bright colors?

Biologists do not fully understand why tropical-water fish, particularly those in shallow seas and from coral reefs, display such a dazzling mixture of patterns and colors. But it is a fact that, as you go from the cold seas around the North and South Poles, through cool seas, and toward the tropical seas near the Equator, the fish become more and more brightly colored.

The colors are caused by a patchwork of color-filled cells in the fish skin. Each one of these is filled with brilliant natural pigments. Some cells make dazzling irridescent colors by the same process that makes an oil film on water show up as brightly colored bands.

The gaudy hues of some individual tropical fish can be explained as different kinds of signals to other animals. The strange-shaped lion fish, for example, has a striped body and fins which are a warning coloration. This fish has extremely poisonous spines on its fins, and other fish species learn to associate the danger of this fish with its patterns. Lion fish are consequently not much bothered by predators!

The tropical clownfish are also brightly colored, probably in a warning way. But they themselves are not poisonous. Instead, they live in amongst the stinging tentacles of sea anemones, to which they are immune. Their warning stripes remind other fish that an attack will involve braving the anemone's tentacles.

The bright shades of many other tropical fish are of use in enabling males and females of the same species to recognize each other as mates. There is a greater need for this in the tropics than in cooler waters, since warm seas have a far greater number of different sorts of fish in them. More sorts (species) means more possible confusion about mating partners.

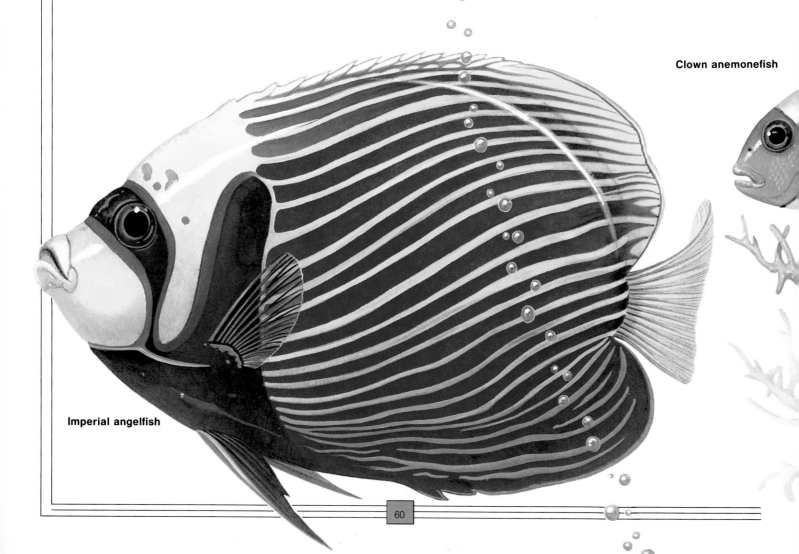

Clown anemonefish

Imperial angelfish

What food can we get from the sea?

Today, all around the world, fish of all kinds, squids, octopus, shrimps, prawns, lobsters, sea urchins, mussels, oysters, and other shellfish are captured or gathered from the sea to feed people. All fish and shellfish are high-protein foods.

The seashores and the oceans have been important sources of human food for thousands of years. Ancient sites, where people lived many thousands of years ago, still contain huge piles of shells—of mussels, cockles, limpets, and oysters, thousands of years old. Today we fish throughout the oceans, and most of the large catches come from sea-going fishing boats that can travel out into the productive waters of the shallow seas bordering the coasts. These areas contain plenty of nutrients for the plant and animal plankton, and they in turn attract larger fish to feed on them.

Fishing boats capture fish in a number of different ways. Some use drift nets, which are left floating like a curtain in the sea to entangle shoaling fish such as herrings. Long, bag-shaped, open-mouthed trawl nets are dragged along the sea bottom to catch flatfish such as plaice and flounder, as well as cod. Fishing lines may also be used; commercial lines up to 20 miles (32 kilometers) long have thousands of hooks spaced along them.

But there is a price to pay for fishing on a huge commercial scale. Overfishing in the recent past has not only reduced the number of fish being caught today, but has also damaged the ecological balance of nature in several ways. Overfishing of herring, for example, has deprived puffins of their staple food, and they have been forced to eat fish lower in nutrients, such as sand eels. As a result one of the world's largest puffin colonies, off the coast of Norway, produced no young in 1980.

Moorish idol

73

Can the sea be farmed?

Yes—fish farming and the farming of shellfish and crustaceans such as shrimps are becoming important in many parts of the world. Fish farms exist for both freshwater and sea-living fish, but the main species now farmed on a commercial scale are trout, salmon, and eels.

The basic idea behind fish farming is to keep a growing colony of the fish being cultured in some form of large enclosure. This colony is fed either with artificial foodstuffs or from natural food sources, and it can be cropped whenever the fish grow large enough. The advantage of the fish farm over ordinary fishing techniques is that the fish are simply waiting in a tank, pool, or cage to be caught and eaten.

Mussels and oysters are regarded as delicacies, as well as being very nutritious, and they too are farmed, in "beds." Both these animals live in estuaries and on rocky shores that are rich in plankton, but their numbers are limited, in nature, since there are not enough places for them to settle on. The supply can easily be increased by setting out rafts and posts on which young larvae can grow to maturity.

In places as far apart as Wales and Japan, seaweeds are collected and valued as food, as are edible plants such as samphire from estuaries. The seaweeds are also an important source of the natural food additives called alginates, used to thicken milkshakes and soups.

Oyster beds in France

74

How do animals and plants reach islands in the middle of the ocean?

Certain plants and animals have managed to colonize ocean islands hundreds or even thousands of miles from the nearest land. Despite its remoteness, and without human interference, an island becomes covered with its own mixture of plants and animals soon after it rises from the ocean as a growing volcanic mountain.

The plants arrive either as windblown seeds, as floating, salt-resistent fruits, as seedlings carried on floating logs or other plant debris, or as seeds stuck to birds' feathers or carried in their droppings.

Animals such as fish, seals, or seabirds may swim to a remote island; other birds, insects, or bats could fly or be windblown; small animals may float attached to logs.

The coconut palm grows on sandy beaches. It originated in Southeast Asia and produces the huge fruits known as coconuts. Their fibrous husk is buoyant and helps the coconut to float from island to island. When it floats ashore, the large seed inside can germinate and produce a new palm tree.

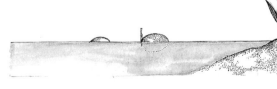

Are island animals the same as mainland ones?

Island animals usually originate on the mainland and start off being the same. In time, though, animals on isolated islands begin to change. These changes are adaptations to their new environment, which is part of the process of evolution.

The original, colonizing animal will split into a range of different, but closely related, types, each one suited to some specialized way of life (or "niche") on the island. This pattern of change is most clearly seen on extremely remote groups of islands. A probable reason for this is that there are no competitors already there. With the island to themselves, the early colonizers from the mainland could begin to adapt and diversify, without worrying about competition.

The two groups of islands in which this story is played out in its clearest fashion are the Galapagos Islands and the chain of Hawaiian Islands—both far out in the Pacific. In both cases, a finchlike bird from the American mainland managed to get to the group of islands in large enough numbers to start breeding. The original species was different in each of the groups of islands, and in both cases the original species has since split into an extraordinary range of different bird types.

On the Galapagos Islands the colonizer birds became known as Darwin's finches because they were the collection of species that the naturalist Charles Darwin saw during his visit on the research ship "HMS Beagle." Together with the Galapagos tortoises—a different form on each island in the group—these birds helped Darwin to develop the idea of evolution itself.

The Hawaiian Islands are probably the most remote group of sizeable islands in the world. The pattern of evolutionary change shown so well by the honeycreepers is also shown by other groups of Hawaiian organisms, including many of the plant species of the islands, and several insects, including fruit flies.

On the Hawaiian Islands the finch ancestor evolved into at least 23 different species of "honeycreepers", living on different islands in the group or in different parts of the same island. They now have a remarkably wide range of beak shapes, adapted for the different sources of food that they eat. Some still have the short, stubby seed-eating bill of the finch ancestor. Others have long, thin, curved insect- or nectar-feeding bills to reach into deep, tubular flowers. Another uses plant spines as a "tool" to winkle out insects from crevises.

Of the original 23 species of honeycreepers, eight are now extinct, having over-specialized in their feeding habits. The surviving species are more generalized feeders but still display an astonishing range of adaptations.

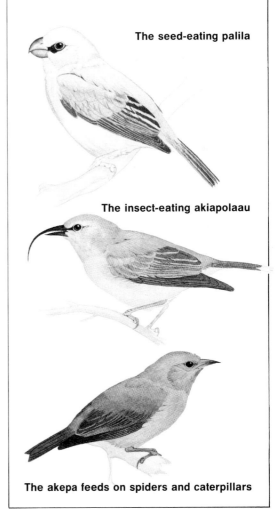

The seed-eating palila

The insect-eating akiapolaau

The akepa feeds on spiders and caterpillars

76

Do dead plants just disappear?

The spectacular fruiting body of a fungus on the floor of a rain forest. Toadstools and other fungi are the reproductive bodies of plant-consuming organisms called hyphae. They release thousands of microscopic spores which can germinate into new masses of hyphae on every dead plant.

When a plant dies it quickly loses most of its water and becomes withered, then turns yellow or brown. Sooner or later, a dead plant is broken down with the help of small creatures known as "detritus feeders," such as insects and worms, and decomposing organisms such as bacteria and fungi.

When you walk in the woods, you can see plant-eating insects and worms once you turn over the leaf litter (the carpet of dead leaves that covers the ground) with your foot. You will also see toadstools and other fungi sprouting from the ground and from dead wood. Besides these visible organisms are microscopic yeasts and bacteria that you cannot see, consuming and breaking down dead plants.

The insects and worms eat and digest fallen leaves, dead plants, and wood fragments. The fungi send thin, cottonlike threads called hyphae into the dead plants. These threads can digest the remains and break them down.

77

What happens to the bodies of dead animals?

They are eaten and broken down like those of dead plants. But there are many larger, specialized animals that eat carrion—dead animals—and they reach the remains before the smaller decomposers.

In most areas of the world there are large, high-soaring birds that stay in the warm air currents watching for signs of an animal carcass. When they see one they drop down to devour it with their strong, ripping beaks. In temperate areas, these scavengers are birds such as buzzards or bald eagles. In South America the bird might be a condor, and in the plains of East Africa, one of a number of species of vulture.

Other carrion-eating animals are found in Africa. After a kill by a lion, leopard, cheetah, or hunting dogs, a number of animal species besides the vultures are waiting to eat anything left behind after the predator has finished. These include jackals and hyenas. A hyena's jaws are powerful enough to crunch up the bones, leaving only the animal's horns. But there is a specialized decomposer, a small moth, that lays her eggs on the horns. The caterpillars that hatch from them can eat and digest horn!

What is a parasite?

A parasite is any organism that lives in or on another animal or plant (called a host), causing it to be damaged or diseased.

All the microbes that cause infectious diseases—viruses, fungi, bacteria—are called microparasites, since you need a microscope to see them. Macroparasites—the larger parasites—are worms of various types (flukes, tapeworms, and roundworms), leeches, and a wide variety of parasitic invertebrates, most of which feed on blood.

There are even parasitic "fish," mammals, and birds. Lampreys are blood-sucking relatives of true fish which live on other fish species. Vampire bats are flying mammals that feed on the blood of birds, mammals, and even humans.

Cuckoos are called brood parasites. They lay their eggs in the nests of other bird species, and the host birds bring up the cuckoo young. The breeding success of the hosts is damaged by the cuckoos, because they rear cuckoo young, not young of their own kind.

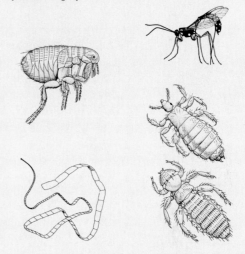

The many types of parasites include a wingless, blood-sucking flea that jumps up to feed on its host (top left); a tapeworm (bottom left) that lives in the gut of its host, absorbing its food ready-digested; an ichneumon fly whose larvae eat other insects alive from within (top right); a blood-sucking human body louse (middle right); and a feather louse which can eat feathers (bottom right).

A jackal and three hyenas compete for the bones and hide of a killed giraffe.

79

How do animals live together?

Although many types of animals live their lives almost completely alone, apart from a short mating period in the breeding season, others live together in groups for part of or all their lives.

The smallest groups are family groups. A mother animal may stay with her young for some time, teaching them survival tricks and how to get food. In long-lived animals such as bears or tigers this family group may stay together for many years.

Larger groups of so-called "social animals" are extended family clusters, with adults and young living together. Lion prides, baboon troops, and packs of hunting dogs are examples of these.

Among the insects, there are social species that form enormous groups all based on the offspring of a single huge female. These groups may contain hundreds of thousands or even millions of insects. Termites, ants, wasps, and bees usually live in "societies" like this, and they all help and support each other.

Certain animals will form huge groups but only at difficult times of year. Starlings do this in winter, with flocks of more than a million birds roosting together. The herbivores of the African plains, including wildebeest and zebra, make up large herds which move together to migrate.

The society of leaf-cutting ants includes workers, and "soldiers" that guard the nest.

80

Do living things ever help each other?

There are many examples of animals, plants, and microbes of different species living closely together in a relationship where they depend on each other. This type of mutually helpful partnership is called "symbiosis" or "mutualism." The advantage gained by each of the partners in the relationship can concern getting food, gaining protection or shelter, or improving the chances of breeding successfully.

The partnership may be between an animal and minute microbes. Termites are social insects of the tropics that can digest the cellulose in dead wood. They can do this only because their gut contains a collection of bacteria and single-celled animals called flagellates. It is actually these tiny partners that do the digesting. In return, they themselves get shelter and a constant supply of food.

Helpful, symbiotic partnerships between animals of different species are common in all types of environment.

In the sea, several species of hermit crabs live in the discarded shells of sea snails. These shells often have sea anemones living on them. The hermit crab gains protection from larger predatory crabs and large fish, which are discouraged by the anemone's stinging tentacles. The anemone, which cannot travel far without help, is carried into new areas for feeding, and also probably eats scraps left behind from the crab's meals.

81

Which animals pollinate flowers?

Flying insects, crawling insects, and many species of birds and bats all pollinate flowers. They are attracted in the first instance by the color, patterning, or scent of an open flower.

Of the insects species that do this, it is probably the social bees and wasps, as well as butterflies and moths, which play the greatest role in carrying pollen between flowers.

Among the pollinating birds, tropical groups such as the hummingbirds and sunbirds are most often involved in pollination. Tropical flowers are brightly hued and the plants that produce them are strongly constructed to take a perching bird's weight. The flowers open during the day and make a large quantity of sweet nectar.

The advantages to the flowering plants are great. They have pollen passed between plants for the

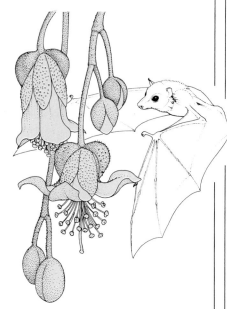

Bats pollinate flowers that open at night. They feed on the nectar and carry the pollen away on their furry heads and necks.

purposes of reproduction, without having to rely on the wind to blow it. In the rain forests, where there is little air movement at the lower levels, most of the flowers are bird- or insect-pollinated.

82

Why do they do it?

On the African plains, birds called oxpeckers live their lives in close association with large herbivores. They scuttle over their skin, removing and eating blood-sucking flies and ticks.

The pollinating animals gain two types of food from their visits to the flowers. The first is nectar—a sweet, nutritious juice manufactured by the flower in structures, often pocketshaped, called nectaries. The second is pollen itself, which can be used as food by bees and a few butterflies.

In gathering these foods, the animals become dusted or covered in pollen. When they move to the next flower, some of it may well end up close to the female part or "ovule" of the flower, which becomes a seed when pollinated.

Pollinating animals often

have special methods for removing foods from flowers. Bees have pollen "baskets" on their legs. Some moths, such as the hawk moths, have very long tongues for getting nectar at the ends of long tubular flowers such as orchids or honeysuckle.

The exotic shapes of some orchids are designed to ensure pollination by bees. A South American orchid produces an intoxicating substance which attracts the bees. An "intoxicated" bee tumbles into the flower, and the flower's shape guides its fall, so that it collects pollen and also deposits pollen it has carried in.

83

How do animals kill their prey?

Animals that kill other creatures for food use a wide variety of different methods. Nature is said to be "red in tooth and claw"—meaning that many animals live by killing and eating others—but teeth and claws are just two of the attack weapons that can be used. Depending on the animal, they may use beaks, jaws, webs, poison, or crushing power.

Before they kill other animals for food, though, hunting animals have to catch their prey. They may do this alone or as part of a hunting group. Wolves, hunting dogs, and lions use the group method. Their tactics are those of "chase" or "wait."

Chasing animals use speed to catch their prey. Sharks can swim faster than the fish they eat, peregrines can dive and maneuver faster than the birds they capture on the wing, and a cheetah can outpace any other land animal.

The waiting animals sit, usually well camouflaged, and wait for their prey to come near. Then they attack. Anglerfish and chameleons use this method. In the case of web-building spiders, the prey insects actually trap themselves by flying or stumbling into the sticky silk web. Having constructed the web, the spider waits nearby for a meal to be caught.

The kill of a predator may be made by a bite with strong teeth, as used by a lion, or with a sharp, strong beak such as that of an eagle. The anaconda snake crushes its victim within the coils of its strong and muscular body, while rattlesnakes, octopuses, and jellyfish, all use poison to paralyze and kill their prey.

The brown pelican hunts for fish food in the seas off the American continent. It cruises above the water until it spots a shoal of fish near the surface. It then goes into a dive, angling back its wings at the moment of contact with the sea, and plunging just below the water surface. As it submerges, it traps fish in its huge beak. "A wonderful bird is the pelican—its beak holds more than its belly can!"

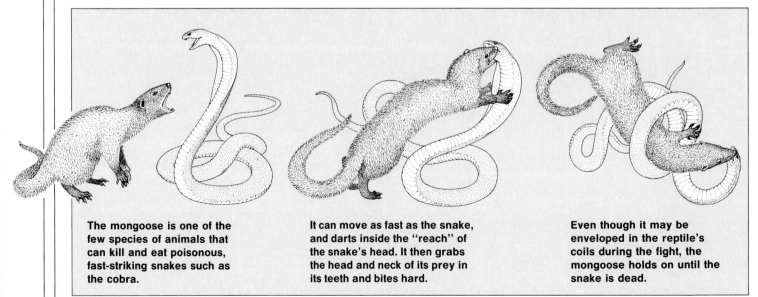

The mongoose is one of the few species of animals that can kill and eat poisonous, fast-striking snakes such as the cobra.

It can move as fast as the snake, and darts inside the "reach" of the snake's head. It then grabs the head and neck of its prey in its teeth and bites hard.

Even though it may be enveloped in the reptile's coils during the fight, the mongoose holds on until the snake is dead.

How do animals avoid being killed?

Prey animals have developed many ways of increasing their chances of survival. Most important are their powers of observation, which enable them to become aware of the presence of a hunter before it sees them. Once a predator has been spotted, the prey animal has several possible strategies: it may run away, protect itself by means of armor, fight back, or hide.

Many animals that are commonly eaten as prey have very well-developed senses: they have large ears, good eyesight, and sensitive noses. They keep watch for the presence of hunters almost continually.

Rabbits and hares are good examples. They are regularly chased and eaten by many different hunters, from golden eagles to foxes, so they are constantly on the alert. With their huge ears and large eyes they survey the world around them for possible threats. If danger threatens, their strong hind legs give them a superb sprint start. Deer, antelopes, and horses also use their long, powerful legs to escape from killers.

Armadilloes and hedgehogs use physical protection, in the form of armor plates or spines, to defend themselves against hunters. Skunks produce a smelly chemical that can be squirted at an attacker. The teeth of a hippo and the tusks of an elephant are weapons that prove more than a match for most predatory animals!

Many commonly hunted animals have developed sophisticated camouflage techniques to avoid detection.

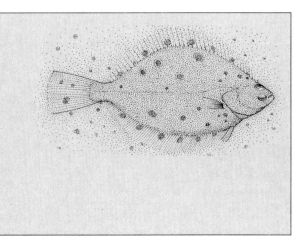

Most camouflage methods depend on the match of an animal's color or its pattern to the background, or on breaking up an animal's distinctive body outlines. All of these techniques make the prey virtually invisible to its predator, so there is no need to run away or fight.
The plaice, like other flatfish, lives on the sea bed. To avoid detection, it first shuffles its fins so that they are partly buried in the mud, sand, or gravel at the bottom of the sea. The color cells in its skin then change their sizes so as to match closely the color and graininess of the sea bed.

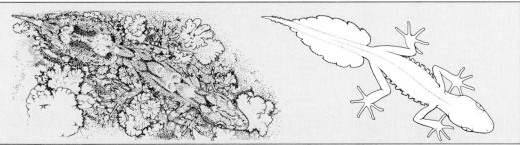

A Malaysian flying gecko blends into its background (right) because of its frilled edges. The fringes on the sides of the body and tail (far right) allow it to flatten out.

A normal mantis stands out against its background (right). But camouflaged species (far right) have wing covers which, in color, shape, and veining, look exactly like leaves.

85

Have people always hunted animals?

Early humans hunted and killed for food.

There is no doubt that humans have always eaten animals as well as plants, so they must have hunted and killed them first. Scientists know that the human species evolved over two million years ago, and alongside the fossils of these early human beings have been found the bones of the animals that they had killed and eaten. Until about ten thousand years ago, humans lived as "hunter-gatherers," killing and eating wild animals as well as picking fruits, leaves, and berries, and digging for roots.

It is possible that one of the early reasons for the success of the human species was its ability to eat almost anything. We ate all the types of animals that we could collect or capture—birds, mammals, reptiles, insects, fish, snails, and many others. We also consumed every type of vegetable food. Because they were omnivorous, early people could survive in almost any habitat.

86

Why did humans first tame animals?

In order to make use of them! Groups of people around the world began taming wild animals some ten thousand years ago, or even earlier.

It is easy to imagine how this might have first happened. A human hunting party out killing for food would have found it easy to capture, alive, the young of any adult animals, particularly mammals, they had killed.

The first animal to become tamed, or domesticated, in this way was probably the dog. A young wild pup taken from its dead mother at an early age is quite easily tamed. If it were kept in a human group and fed, it would rapidly become used to the presence of people in the same way that, in the wild, it would become part of a dog pack.

As a group-hunting animal itself, a dog is used to cooperating with its partners in catching food. It is likely that tamed dogs brought up with people showed themselves to be useful as hunting companions. With their good noses, speed over the ground, and ability to run down and kill many types of prey, dogs would have been invaluable helpers. They would certainly have earned the small amount of food that was necessary to keep them.

Later, when animals were domesticated on a larger scale (see Question 87), dogs would have become useful in the way that sheepdogs are today, to control half-tamed flocks and herds of food animals.

Larger mammals such as horses, buffaloes, donkeys, oxen, and llamas were tamed partly to be used as "beasts of burden." They could transport people over long distances, or carry other heavy loads for them. With the beginnings of farming, their strength was put to other uses, in particular for pulling the primitive plows used to break up the soil before planting crops.

Both the animals killed for food, and the carrying and pulling animals, were also an important source of clothing materials, especially wool and leather.

The beasts of burden—camels trekking across the desert in Ethiopia, laden with salt.

What animals do we use as food?

Soon after dogs were tamed, human societies in all parts of the world began to domesticate certain animals as meat and egg producers. These were mainly sheep, goats, cows, pigs, and chickens.

Sheep, goats, and cows are all herbivores. They eat the tough, dry vegetable materials that people cannot digest. Keeping these animals made a high-protein, nutritious food much more easily available than going out and hunting for it. It also made more efficient use of the habitat's food potential than was possible before, since this meat protein was being produced indirectly from plants which people themselves could not eat.

Pigs and chickens were also tamed. The pig is an omnivore that can find food in many different types of surroundings, and can eat scraps and refuse from a human camp. Chickens, which were probably developed from the red jungle fowl of Southeast Asia, can feed on almost any food too. They were a good source of meat and eggs, and were easily transported when people moved around.

Male and female domesticated fowls

88

Why do creatures die out?

They die out because they stop being successful enough at breeding. In other words, the adults die at a faster rate than their young are being born. This breeding failure can be due to a change of climate which makes life difficult for the species, the arrival of new, more successful competitor species, the loss of an important food source, or of their whole habitat, or even human predation.

All creatures die out, or become extinct, in the end. Looked at over a long enough time scale, every species develops, survives for a period of time, then dies out. The period of survival can be anything from a few hundred thousand years to hundreds of millions of years.

Scientists have puzzled over the extinction of all the dinosaur species, which was completed about 60 million years ago. For 130 million years before that, these huge reptiles ruled the Earth.

Many reasons have been suggested. Was it a climate change? Was it increasing competition from the warm-blooded mammals? Or was it a global disaster caused by a huge meteorite falling to Earth which produced a drastic cooling of the planet's temperature? Nobody knows the true answer.

Pterodactyls are extinct prehistoric flying reptiles that belonged to a group called the pterosaurs. They had membranelike wings which, like those of today's bats, stretched behind very long arms and fingers. They could probably only glide, rather than fly by flapping their wings. They all died out over 60 million years ago.

89

Have any animals become extinct recently?

In the last few hundred years animals of all sorts have been dying out increasingly rapidly. The terrible but simple reason for this is that there are too many people on Earth to give all the other animal species enough room to exist.

As human populations have continued to grow (there are now over five billion of us crammed onto the Earth), we have killed more and more animals for food. Perhaps more importantly, we have taken their habitats from them, by turning the natural ecosystems of the world into towns, cities, factories, and farmland. When they are killed in excessive numbers, or deprived of the habitats that support them, animal species die out.

Over the last few hundred years the most famous extinctions have been those of the dodo in the Indian Ocean, the passenger pigeon in North America, and Steller's sea cow in the waters of the North Pacific. The dodo, a flightless bird living on the island of Mauritius, was hunted by man, and also suffered when pigs were introduced onto the island, and destroyed their nests.

Steller's sea cow was a large marine mammal which reached 24 ft (7 m) in length and fed purely on seaweed. The sea cow was made extinct only 25 years after its discovery in the eighteenth century.

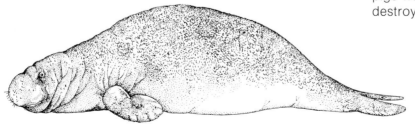

What are the world's most endangered animals?

There are, unfortunately, too many animals to list whose numbers in the wild are steadily declining. Endangered species are those likely to become extinct soon if conservation measures are not taken.

Most of the endangered large animals are in trouble because of human harm. Animals may be directly killed, not only for food but also for their skins (as in the case of crocodiles), their tusks (as with elephants for ivory), or their horns (as in the case of the black rhino). Butterflies have been brought to extinction by over-enthusiastic collectors.

Humans can also cause an animal to become extinct by interfering with its surroundings in some way. Entire rain forests are being destroyed for agriculture, nesting and feeding sites are being disturbed by people, and poisonous wastes are being dumped in rivers and the sea, killing marine creatures.

The tree-living Amazonian sloth has been put at risk by burning and clearing the rain forests for agriculture.

The Californian condor is nearly extinct due to human disturbance in its nesting and feeding areas.

The Mediterranean monk seal is endangered by tourist invasion of its once-secluded sandy breeding beaches.

Gorillas living in the forests of central Africa are down to very low numbers but are still being killed by poachers.

What are the most common animals?

In terms of numbers, the most common animals must be some of the very small ones. Single species of small insects, worms, and snails, or of single-celled animals, can exist in numberless billions of individuals. Just think how many aphids (greenfly) there might be in one field of an infected crop. Or imagine how many houseflies there could be during the summer in one city.

Of the bigger types of animals, our own species can make a good claim to being present in the largest numbers. The human race is distributed all over the planet, and with a total population of over five thousand million people, it is more numerous than any other large mammal on Earth.

Some of the other very numerous species are ones that are closely associated with human societies in one way or another. Among the mammals, rats and mice have reached very high total numbers, probably because of their links with human homes. Among the birds, house sparrows—birds that are always found close to people—are also one of the most numerous sorts.

92

Do zoos and wildlife parks help to save endangered animals?

Yes, they may help to save endangered animals in one of two different ways. Zoos and parks with properly organized aims can sometimes help by building up breeding groups of endangered animals in protected conditions. If they breed successfully like this, they will increase in numbers.

The second way of helping is an indirect one. By enabling people in all parts of the world to see the beautiful creatures that may only live in one remote region of the globe, zoos and wildlife parks are a way of educating city people about the wildlife of our planet. In some cases the zoos can help encourage public support for campaigns to save endangered species, by teaching people about the animals in danger of extinction.

It must be said, though, that on a worldwide scale, zoos probably remove more animals from the wild than they put back through breeding programs. Many of the world's zoos are still mainly places for display and entertainment.

Tourists "on safari" at a game reserve in Kenya will see animals living as they do in the wild.

93

How are animals bred in captivity?

If animals are to breed successfully in captivity, they have to be provided with all the right conditions for normal courtship, normal mating, nest-building, and rearing their young. These conditions differ from species to species but specialists can ensure they are provided.

But despite a great deal of research, it is still difficult or even impossible to get some endangered animals to breed well in captivity. Giant pandas and dolphins have both proved troublesome. Faced with these difficulties, a number of high-technology approaches have been tried to improve breeding success. Remotely controlled video cameras are used to watch animals without disturbing them in the breeding season, and blood samples are taken to check on levels of sex hormones to pinpoint the best time for breeding. Artificial fertilization methods have even been tried, with the great apes.

To produce healthy offspring, care has to be taken to avoid "in-breeding," that is, mating animals that are closely related. Animals are transported around the world from zoo to zoo for mating attempts.

The giant panda

95

Are animals in zoos and wildlife parks ever deliberately killed?

Sometimes they are. Neither zoos nor wildlife parks are completely natural ecosystems. This means that, on occasions, the breeding of some animal types will produce more young than can easily be housed or fed in a zoo, or supported by the habitat in the park.

In zoos, such surplus animals are usually transported to other zoos or back to the wild. If this is not possible, though, they are sometimes deliberately and humanely killed.

Similarly, in some parks and game reserves, particular species can reach such high numbers that they are a threat to their surroundings. In some African reserves, for instance, elephants have reached damaging population levels. There are so many of them that their feeding is destroying the vegetation cover. In such circumstances, controlled killing programs, known as culling, are sometimes used. This has happened in recent years in the Tsavo game reserve in Kenya.

94

Are they ever released back into the wild?

Yes, sometimes. There are a few clear success stories, in which endangered animals have been bred in captivity and then reintroduced into natural conditions to start up healthy populations in the wild. Examples include the white rhino and the Arabian oryx.

One of the best examples is the story of the near-extinction and recovery of the ne ne or Hawaiian goose. The ne ne was brought to the verge of extinction by 1950, because it was hunted by both humans and its predators, and through the destruction of its natural habitat. Only 25 pairs of adult breeding birds then existed.

At that desperate stage, a small number of birds were taken, as a last resort, to the Wildfowl Trust's reserve at Slimbridge on the banks of the river Severn in England. There, and also in captivity in Hawaii, the birds began to breed. The ne ne eggs were put under broody chickens for hatching, to increase their chances.

In this way, good stocks of the birds were built up. Some of these were then sent back to Hawaii to set up populations in protected reserves on remote islands.

The ne ne is restricted to the Hawaiian islands in the middle of the Pacific Ocean. It seems to have evolved from a stock of ancestors similar to Canada geese, which were blown off course during their migration. The ne ne is now quite a distinct species. It is much smaller than the Canada goose, with proportionally smaller wings and less webbed feet. It does not migrate.

Can the whales be saved?

They can be, if commercial whaling is stopped for long enough. Although they are by no means the only endangered sea animals, whales have become symbols of endangered animal life in the seas. Particular species of dugongs, manatees, seals, and turtles are also dangerously close to extinction. Some fish stocks have fallen to very low levels too.

Several types of whale, including the humpback whale shown below, are seriously declining in numbers due to previously high levels of whaling kills. Other endangered species are the gray whale, the right whale (so-called because it was the "right" type for hunting), and the blue whale. The case of the blue whale, in particular, shows just how damaging human intervention can be.

The blue whale is the largest mammal that has ever lived. It can weigh more than 140 tons and grow to a length of more than 100 feet (30 meters). Despite its huge size, it feeds solely on small planktonic crustaceans called krill, which it strains out of the sea with the whalebone "baleen" plates in its mouth (see Question 65). It can eat over four tons of these shrimps, or krill, in a single day.

But this awesome giant of the oceans, which has a long life and a slow breeding rate, is hunted by humans for its meat and its oil. As a result of excessive hunting it has been reduced to very low numbers, perhaps already too low for survival. The species has been completely protected from commercial whaling since 1967, but it remains to be seen whether this protection will allow it to recover.

The Japanese whaling fleet is still killing baleen whales. If this is not stopped, it is likely that the smaller minke and sei whales, which they now hunt, will soon be on the endangered list too.

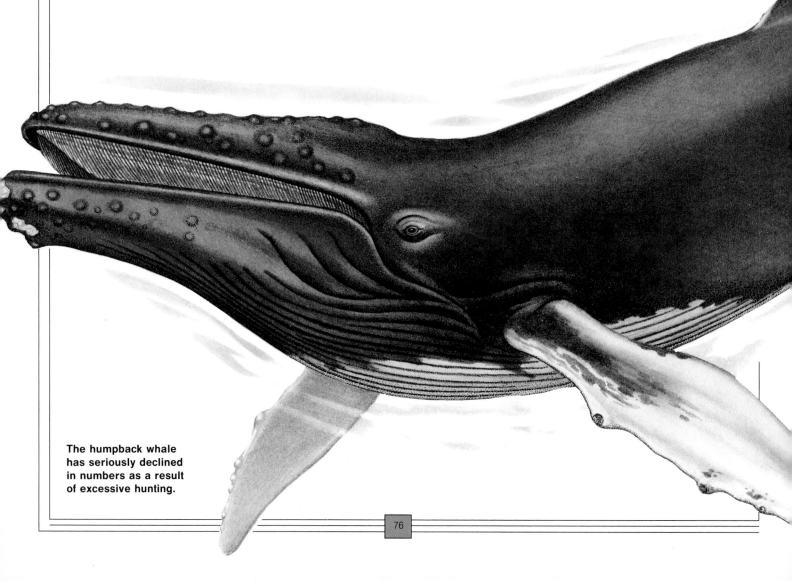

The humpback whale has seriously declined in numbers as a result of excessive hunting.

97

Why have seals been dying in large numbers in the seas off Europe?

There are several reasons why in 1988, seal deaths increased to alarming levels in the seas off Scandinavia and northern Europe. The reasons show the pressures on these vulnerable sea creatures in an industrialized world.

Some of the seal deaths were caused by natural toxins, or poisons. These poisons are produced when particular types of marine plants, or algae, grow in such large numbers that they form an algal "bloom" in the sea. In the Baltic Sea, this bloom was brought about by the run-off of artificial fertilizers from farming land into rivers, and from there into the sea.

The nutrients in the fertilizer probably triggered off the massive growth spurt of the algae.

Another reason for the unusual levels of seal deaths was due to a virus disease similar to the virus that produces the disease called distemper in dogs. It is likely that pollution in the ocean has reduced the seals' immune defenses against the new disease. It is also possible that human influence played a part in a different way, since a mass of carcases of dogs that died from a distemper epidemic in Greenland were dumped in the sea. This may have infected the seals directly.

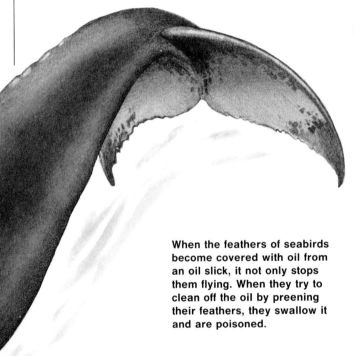

When the feathers of seabirds become covered with oil from an oil slick, it not only stops them flying. When they try to clean off the oil by preening their feathers, they swallow it and are poisoned.

98

How does pollution harm sea creatures?

It harms them by damaging their bodies, either outside or inside. There are many different types of serious marine pollution, and the important groups of pollutants include: oil, heavy metals, pesticides, fertilizers (see Question 97), sewage, and radioactive wastes. Pollution levels are worst in the shallow seas off the coasts of industrialized countries.

Oil from oil-rig accidents—oil tankers that are damaged, or sink, or which illegally flush out their tanks at sea—all form a floating pollution hazard known as an oil slick. These slicks are eventually dispersed and broken down by microbes in the sea, but not before the feathers of

swimming seabirds are damaged.

Thousands of tons of heavy metal salts (such as those of mercury, zinc, and lead) pass into the sea every year, down any river system that has major industries along its banks. The river Rhine in Germany discharges its pollutants into the North Sea, for example. These substances, and similar ones from the anti-fouling paints on ships, are dangerous to almost all sea life. They stop certain chemicals in their bodies from working.

Pesticides such as powerful insecticides reach the sea from farming land, via rivers. These substances, too, can poison many types of marine life.

How can endangered animals be protected?

There are many things that can be done to save animals which are in danger of immediate extinction. These can be divided into short-term "emergency" measures, and longer-term assistance.

The quick-acting, "trouble-shooting" methods of protection are designed to stop a species dying out completely in the immediate future. These include breeding captive populations of the animals in zoos and parks (see Question 92), and making laws against the unnecessary killing of endangered species, such as the blue whale (see Question 96). The laws must, of course, be policed effectively, too.

More long-sighted plans must find ways of making sure that the endangered species, having been saved from immediate extinction, can continue to survive indefinitely. These plans involve ensuring that enough proper natural habitats exist for these animals to live in. This can only be done by careful laws concerning destruction and pollution of all types of environment.

It is most important that ways are found to preserve big enough stretches of the right habitat in a natural state for the animals to live in. For instance, the bamboo-eating giant panda will remain alive on Earth only if enough large areas of bamboo forest are preserved in mainland China.

Mountain gorilla

Birdwing butterfly

Black caiman

Helmeted hornbill

Orangutan

The illustration shows just a few of the many animal species now under threat. The orangutan, mountain gorilla, and Sumatran rhinoceros all need deep cover and a lot of space. Their life cycle is disrupted and their breeding rate slows down if their natural habitat is disturbed, as happens when areas of forest are cleared for agriculture.

Imperial parrot

Sumatran rhinoceros

White-winged wood duck

100

Are any insects endangered?

Probably hundreds of species of insects become endangered and extinct without us even knowing about it. This is because there are many thousands of insects we have not discovered yet, as well as the million and more that are known to be in existence.

In areas of the world such as rain forests, where habitats are being destroyed, the insect types which depend on that habitat can vanish from the face of the Earth before humankind has named them. Their dependence on the rain forest habitat is usually tightly linked to particular plants. Many insects feed only on the leaves or seeds of an individual plant type and, if that plant vanishes because of forest clearance, the insect will disappear too.

More direct human pressure can put insects at risk of extinction. Beautiful butterflies and moths from the tropics are highly prized as collectors' items. The larger and rarer they are, the more money they fetch. A single specimen of the marvellous birdwing butterflies of Papua New Guinea, for instance, is worth hundreds of dollars. This type of hunting pressure can soon lead to insects becoming endangered.

The large blue butterfly (far left) has become extinct in Great Britain during the 1980s. Its loss has been triggered by the disappearance of a species of ant on which it depended for food and protection.

The giant earwig (left) has recently reached the brink of extinction. It lived on the volcanic island of St. Helena in the South Atlantic, but none have been seen since the 1960s. Human colonization of the island resulted in the destruction of its forest cover, and the introduction of alien species such as rats and giant centipedes threatened it further.

101

Are insecticides useful or harmful?

They can be both useful and harmful. It depends which insecticide you choose and where and how much of it you use.

Insecticides are chemicals, either synthetic or natural, used to kill pest insects. These include biting insects such as mosquitoes, which spread diseases such as malaria, and insects that eat plants or infect them with a disease. Locusts come into the first category, and aphids (greenfly) into the second. Most insecticides are used by farmers to protect their crops or by health authorities to control killer diseases such as malaria.

The problems linked with insecticides occur because these poisonous substances may kill both the pest insects that are the target and also other creatures that are not. If you spray them on a field, you may certainly kill the aphids, but unfortunately you may also kill pollinating bees and butterflies, useful insects that keep other pests under control, fish in nearby rivers, and birds that eat the poisoned insects. Several insecticides not easily broken down in the soil are also poisonous to humans.

The use of some synthetic insecticides such as DDT is either banned or very restricted because of the dangers associated with their use. But it is now possible to use more selective and non-dangerous insecticides that kill only the target insects. Some of these, such as the pyrethroids, are based on natural substances obtained from plants.

Why are insect pests so difficult to get rid of?

A locust

Individual pest insects are not difficult to kill. It is perfectly easy to use pesticides to kill a whole range of harmful insects, such as houseflies, tse tse flies, mosquitoes, blackflies, and sandflies, that spread human diseases. They can also be used to kill crop-damaging insects such as locusts, aphids, scale insects, and weevils. But the difficulty lies in stopping the insects breeding faster than they can be killed.

Insects have phenomenal breeding rates. They lay large numbers of eggs and, being small animals, can grow from eggs, through the larval stages, to egg-laying adults very quickly—sometimes in just a few days. Their population size, if food is available, can grow staggeringly fast in a sort of "cascade" of breeding.

Not counting casualties, a female fly from a species that reaches maturity in seven days and lays a thousand eggs in its lifetime will produce over 50 billion offspring in one month. Such rates of increase are never actually attained because of natural deaths, starvation, and disease among insects. But the calculation shows just how persistent and careful an insect control program must be to stand a chance against such rapidly breeding foes.

It is unlikely that any pest insect has ever been completely eradicated by human attempts to control it. The best that can be hoped for is that pests can be kept at reasonably acceptable levels by the continual use of insecticides. Some degree of success has also been achieved by introducing natural bacterial and virus insect diseases as biological control agents, and by destroying the insects' favored living and breeding sites.

The locust invasion in Ethiopia—the adult swarm is being attacked with poison bait.

103

Have we made new places for animals to live?

Yes we have. The human impact on animals is usually thought of as harmful, but there is another side to the story.

People have tamed animals such as dogs, cats, sheep, horses, chickens, and cows for thousands of years (see Question 86). We have taken great care to provide good places for these animals to live in, with plenty of food, in our own surroundings. For other, wild sorts of animals, though, the animals themselves—without any prompting from us—have decided that the farms, villages, towns, and cities that we build are ideal homes for them too.

Some of these creatures, such as rats, we regard as pests, since they may carry diseases or eat our stored food. Others, for example the house martins that nest in the eaves of our houses, are considered a sign of good luck and a happy home.

Whether they are pests or welcome guests, there appear to be several reasons why certain species of wild animals come to live alongside humans. Firstly, our buildings in both town and country offer them shelter, and often warmth. Secondly, the areas where humans live provide nourishment for those animals that eat a wide range of different foods (omnivores).

104

What animals live in towns and cities?

A wide range of animals now lives in towns and cities. Behind walls, under floors, in sewers, and in ruined buildings you will find house mice, rats, and scavenging insects such as cockroaches. They feed on our scraps and stored food.

Out in the open, but close to the places where we live, are the birds that have specialized in living near people. Huge numbers of house sparrows live in towns and cities; chimney swifts and house martins nest in our buildings; and flocks of a million starlings at a time roost on window ledges and roofs. In the squares and streets of almost all cities, pigeons walk amongst people, looking for food. And in the country swallows and barn owls nest in barns and outbuildings.

Red foxes and raccoons are more recent converts to city life. They are omnivorous and intelligent animals, and search the city suburbs for food.

Raccoons have adapted well to life in the suburbs of North America. They are mainly active at night and eat a wide variety of foods. Many families put out milk and scraps for them, but they also scavenge through garbage cans.

How do they manage to survive there?

They do so by being adaptable and taking advantage of the special conditions and opportunities that exist in built-up landscapes.

Nesting sites are a good example. The ancestors of birds such as town pigeons and house martins must have made their nests on cliff ledges or under the overhangs of cliffs. Once people started building houses, the birds saw the walls and eaves of these constructions as new, easy-to-use cliffs.

The ancestors of house sparrows probably made their large, almost communal nests in hollow trees or in crevises in cliffs. The roof spaces of houses provided a good imitation of these nest sites. The added advantage was that there was usually plenty to eat nearby.

This extra food source is crucial in inducing many animals to stay close to us. In rich, industrialized societies, with food to spare, people will actually deliberately leave food out for wild birds on a garden bird table, or feed squirrels, pigeons, and ducks in a local park. This food supply, in addition to the artificial warmth of cities, and the shelter offered by the buildings, makes hard winters easier to endure for city-living wild animals.

Manhattan, with its high density of buildings, may seem a hostile environment but it offers homes to several city-dwelling animals.

A fox searches through the garbage for food.

106

What is acid rain?

Acid rain is caused by the sulfur dioxide in the smoke produced by burning coal and oil. Power stations are one of the main sources of the smoke from burning these fossil fuels. In the atmosphere, the sulfur dioxide changes normal rain into rain which is acidic.

Acid rain is rain that has been changed by some of the polluting substances in the air. These substances have been pushed into the atmosphere of our planet by humans. These "pollutants" make the rain acidic, and the acidity in it can cause a number of harmful effects when the rain falls to the ground.

The polluting substances that produce acid rain are mainly those created when "fossil fuels," such as oil and coal, are burned in large quantities. Both coal and oil contain the chemical sulfur, which, when burned, makes a substance called sulfur dioxide, an acrid gas that is carried into the air along with the smoke from the burned fuel.

In the atmosphere, the sulfur dioxide dissolves in the tiny water droplets that will eventually merge together to make raindrops in clouds. Dissolved in water, it forms sulfurous acid, and much of this is turned by the oxygen in the air into sulfuric acid.

This chain of reactions means that the polluting gas from burning coal or oil has turned a harmless raindrop into diluted acid. When this falls on the ground, on trees, or into lakes, it damages the whole balance of nature.

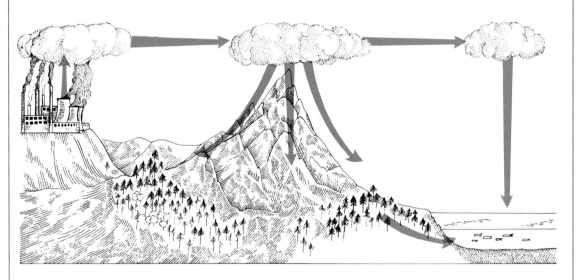

107

Why is it so harmful?

Acid rain is considered harmful because it is damaging the countryside in industrialized lands. The evidence, and the reasons for this airborne pollution, have become clear only in the last ten years.

The clearest evidence for this harm is to be found in lakes in some mountainous areas. In mountain lakes in parts of Scotland, Scandinavia, America, and Canada, the underlying rocks are not able to neutralize or "buffer" the acids, as rocks can in other parts of the world.

So when large amounts of polluted acid rain fall into these streams and lakes, the rain turns the water in the lakes and streams acid. The water then becomes poisonous to various forms of life in it, particularly upland fish such as trout. In addition to damaging the trout directly, the acidity of the water causes dangerous levels of some dissolved metals to drain or leach out from the bedrock into the water. These are also harmful to the fish. There are now many examples of mountain lakes where all the trout are dead.

Acid rain also appears to harm plant life. Many scientists think that acid rain, together with other pollutants, has caused coniferous forests to die back in the last few years. This effect, where the growing tips of the conifers die, has happened in the Black Forest in Germany, as well as in forests in Scandinavia and North America.

This damage could be due directly to the acidic rain falling on the trees' growing points, or could be an indirect result of the changes brought about in the soil on which it falls. The trees could be taking up polluted water and minerals through their roots.

108

What can be done to stop acid rain?

It is possible to neutralize the acidity in a small polluted lake by chemical means, to bring it back to life. But this is only a fairly small-scale solution to an increasingly large threat. The only real answer to the problem of acid rain lies in tackling it at its source. This means controlling the burning of huge amounts of fossil fuels, particularly in the industrialized countries of the world.

In Britain, for example, it is likely that a large proportion of the acid-rain pollution across the country comes from the chimneys of power stations. They burn millions of tons of coal a year, all making

sulfur dioxide. The technology exists for "scrubbing" the smoke and gases emitted from these chimneys to remove nearly all the acidic gas, but it is expensive to install the equipment.

Nevertheless, the authorities that control the industry are now embarking on a program of fitting efficient "scrubbers" to its power stations.

This type of action is obviously required on a large scale in all industrialized countries in order to ensure that the acid rain problem does not get slowly worse, and eventually poison more and more living things.

109

What causes droughts and can they be stopped?

Very long periods of dry weather without any rain cause droughts. This term is applied only to ecosystems, such as grasslands, where the weather conditions are normally fairly wet. In desert lands this lack of rain is expected.

Some droughts are caused by unpredictable alterations in climate patterns. Others appear to happen at almost regular intervals, such as the droughts that have occurred in the Great Plains region of North America over the last two centuries. Scientists are only just

beginning to understand the interactions of the sun, the atmosphere, and the oceans that drive these "cycles" along.

Natural droughts cannot really be stopped in any way. But some scientists think that there is a link between the climatic changes that produce droughts, and the alterations that humans make to the landscape, such as removing the tree cover over large parts of a continent. If this were true, it might be possible to reduce the risk of drought by reafforestation—planting more trees.

The cracked earth of a drought-stricken part of Kenya means that animals will go short of food, and possibly starve to death, if no rain comes.

Are we polluting the atmosphere?

Factories, heavy industry, motor vehicles, and airplanes all produce poisonous gases which are polluting the Earth's atmosphere.

Unfortunately we are, more and more every year, and with an increasingly wide range of polluting substances. The human population has grown to such a huge size that the number of people, and the industries that we develop, produce too many pollutants to be simply absorbed by the atmosphere of our planet.

What is more, the atmosphere is a tiny volume of the Earth's size. From the center of the Earth to the surface where we live, the distance is about 4,000 miles (6,300 kilometers). The "atmosphere" is the gas layer above this. Almost all the gas lies in a layer only some 13 miles (20 kilometers) deep. If you think of the atmosphere as a thin skin around the Earth, it is easy to imagine how pollutants can harm it.

There are several important types of pollution that we introduce into the air. There are acid gases that come from burning coal and oil, and which produce the hazard of acid rain (see Question 106). Burning almost anything at all produces the gas called carbon dioxide, and increasing amounts of this in the

atmosphere create the "greenhouse effect" (see Question 111). Gases from aerosol cans are another kind of airborne pollution (see Question 112).

In addition to these three major types of pollution, there are several others. Exhaust fumes from motor vehicles pollute the atmosphere by adding to the carbon dioxide. They are one of the causes of city smogs such as those in Los Angeles. When leaded gasoline is used, motor vehicles and airplanes also pollute the atmosphere and the ground with poisonous lead compounds.

Finally, there is the problem of dangerous radioactive wastes in the atmosphere. Much of it falls to the ground in rain and we consume it in contaminated food plants, or meat and milk from affected animals. Radioactive materials can cause serious medical problems if they get into people's bones. Accidents at nuclear power stations are another source of this type of pollution, shown most clearly by the terrible accident at Chernobyl, in the USSR, in 1987.

111

What is the greenhouse effect?

The "greenhouse effect" is the process that keeps the Earth warm. Carbon dioxide gas in the atmosphere easily lets through light energy from the sun. When this heats up the Earth and is given off as infra-red heat, the carbon dioxide stops it from getting out again. So the Earth is kept warm like the inside of a greenhouse.

Increasing amounts of carbon dioxide in the atmosphere may be causing the Earth to get even warmer. Carbon dioxide is the gas produced when any organic material is burned, and it is also the gas taken up by plants in the process known as photosynthesis. This gas is now present only in small quantities in the air, but the amount is increasing.

Two human activities in particular are causing the carbon dioxide levels to rise. One is the increasing amount of material that we burn, either to clear land or as fuel. The other is the reduction in the area of the Earth's surface covered with plants, as built-up areas spread. Fewer plants mean that less carbon dioxide is used up.

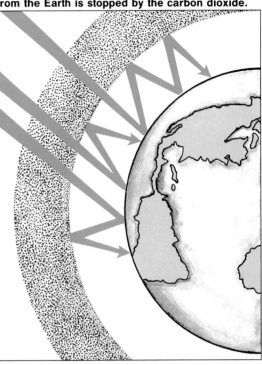

Sunlight energy easily passes through the carbon dioxide in the atmosphere. But heat radiating out from the Earth is stopped by the carbon dioxide.

112

Are aerosol sprays harmful?

Yes, ordinary aerosol cans give out gas every time they are used, which can damage the atmosphere. The gas, called a chlorofluorocarbon propellant (CFC), expands and squirts material out of the can, when the nozzle is pressed.

Unfortunately, it has been realized that this gas, once it reaches the higher zones of the atmosphere, can bring about worrying changes. There it is broken down by energy from the sun and it releases "fluorine atoms." These are able to destroy a gas known as ozone that exists in a region called the ozone layer, about 15 miles (24 kilometers) above the Earth's surface.

The ozone stops cancer-causing ultraviolet radiation from the sun from reaching us down below. If too much ozone is destroyed, the disease cancer could increase. In the past five years, a hole as big as Europe has formed in the ozone layer over Antarctica. Most scientists think that this has been caused by gas from aerosols.

113

What does bio-degradable mean?

If something is biodegradable, it means that it can be broken down by the normal processes of biological decay.

In all types of habitats, there are organisms called decomposers (see Question 76). Their "job" in the ecosystem is to recycle dead animal and plant remains naturally. They do this by eating or digesting the material and turning it into a form in which it may be eaten or used by larger organisms.

The decomposers come in all sizes and types, but the most important are the smallest—the immense numbers of microbes that exist in any soil or in any water system. These include bacteria, fungi, and single-celled organisms called protozoans—all microscopic in size, yet crucial for the recycling of dead materials.

These same decomposers can also break down some of the materials, such as paper and cardboard, made by human activities and industry. Other materials that we make, though, cannot be easily broken down in this way, especially plastics. If they are discarded, they will slowly build up in the environment as unsightly or dangerous pollutants.

In fact, very few materials are completely non-biodegradable. Microbes can eventually break down unusual materials such as mineral oil and rocks.

114

Is it better to package things in paper or plastic?

From the point of view of rotting down naturally, it is usually better to package materials in paper.

Paper and cardboard are biodegradable because they are made from softwood pulp. This product of ground-up tree wood consists mainly of the common carbohydrate molecule called cellulose. Cellulose, when discarded as waste paper and cardboard, is easily and quickly recycled by decomposer organisms. Many fungi and bacteria can break it down and use it again. It is more difficult for them to break down the synthetic molecules of plastics. The decomposers never developed digestive methods for dealing with plastics, a material first made only 50 years ago.

From a conservation viewpoint also, the use of paper or cardboard is to be preferred over that of plastic because softwood plantations are a crop that can be used indefinitely. Provided the trees cut down for paper are replaced by new ones, you can carry on making paper for ever. Plastic, on the other hand, comes from the organic molecules in oil. Oil reserves, once used up, cannot be replaced.

Waste paper, dropped anywhere, rapidly decays, while plastic cartons remain. A walk along almost any seashore in North America or Europe will show this difference in biodegradability. The shores are littered with plastic bottles, plastic bags, nylon fishing lines, and expanded polystyrene packaging.

115

How are materials such as glass and paper recycled?

Recycling, which simply means collecting and using again, can take many forms. As people become increasingly aware that the Earth does not have limitless resources, they see that it is crucial to recycle as much as possible of the materials that we use or make. It is especially important for materials which are scarce, or nonrenewable, or which require large amounts of energy (fuel) in their construction.

Recycling can be carried out on a large scale for dealing with all the wastes of a community. At a waste-disposal factory, this may involve removing all the metals from household waste, magnetically sorting them into different types, and then using them in the manufacture of new metals and alloys (combined metals). Those waste materials that can be burned could be used as the fuel to power the machines at the waste-disposal works.

At a more local level, there are systems for recycling glass and paper that require more initial sorting. Waste glass can be collected, sorted into types of different colors, then melted down and used once again in the manufacture of new glass. Waste paper can be collected, repulped, and used to make new paper. This can be done over and over again with newspapers.

Seashores and riverbanks are often littered with plastic waste.

116

What are the main sources of energy?

The main energy sources we use are fossil fuels, such as oil and coal, natural gas, and peat, all of which are burned to supply energy. These different sources of power are used for heating, generating electricity, driving motor vehicles, and operating machinery. In less industrialized societies, wood is burned in the same way. The other important energy sources are hydroelectric power, and nuclear power.

Each energy source is linked with some problems, either ecological or of some other kind. The fossil fuels are all non-renewable—once we have used them up, there is no way of making them again. But living fuel sources such as wood are at least potentially renewable. All burned fuels, however they are produced, run the risk of increasing the carbon dioxide in the atmosphere, and of creating acid rain.

Hydroelectric power is pollution-free, but it often requires large areas of land to be flooded to produce big enough lakes above the hydroelectric dams. Nuclear power carries the risk of radioactive leaks.

Under study at the moment are nuclear fusion power, wave power, tidal power, wind power, and solar power. It is difficult to predict just what our main power sources will be in the twenty-first century.

117

What good does it do to preserve plants and animals?

All forms of life on Earth depend on one another. We are one of those life forms and we therefore cannot cut ourselves off from other forms of life on our planet. Our own species is just as dependent on green plant life, with its ability to produce energy by photosynthesis, as grasshoppers or zebras are. Like other animals with whom we share the planet, humans ultimately depend for food on the growth of plants.

If we kill insects indiscriminately, for instance, we will be killing off all the species that pollinate our crops. We need to worry about the loss of wild plant species themselves, since their genes (the units of inheritance carried in every living thing) are crucial to the success of plant breeders, who can produce more productive and disease-resistant crops. In the 1920s, when disease threatened to destroy the sugar cane crop in the United States, it was the genes from a wild strain of cane from the forests of Java that enabled a resistant version to be bred.

Amongst the thousands of rain forest plants, many will contain new drugs. If we lose the forests, we lose the chance of finding those drugs.

Our planet's climate, too, depends on properly balanced ecosystems. Destroying them by overgrazing, city building, or forest burning would, in the end, mean that our civilization was produced at a terrible cost: unfarmable land, uncontrolled flooding, and dangerously changed weather systems.

118

How can we feed all the people on Earth and still care for wildlife?

If the human population carries on increasing at its present rate, we will *not* be able to feed all the people and save the wildlife at the same time. This is the harsh reality caused by the fixed size of the Earth's surface, and by the present size of our population—now over five billion people.

Increasing numbers of mouths to feed mean that more land has to be cleared for agriculture and housing. This is beginning to put an intolerable pressure on the remaining large areas of natural habitat and on their wildlife. These undeveloped areas are mainly in the tropics and subtropics. In these areas too are many of the poorest countries in the world, which are also the ones with the fastest-growing populations.

Attempts to save habitats and wildlife must be seen as global problems. They cannot be the sole responsibility of the impoverished developing countries, where the pressures on land and animals living in the wild are most acute.

All the countries of the world must take a share in the responsibility for their conservation. But without some slowing down of the increase in human populations, it is difficult even for ecologists and other specialists to see how this conservation can be achieved.

119

What can ordinary people like me do to help save our planet?

We can all help by putting the world first. Governments and societies will, in the end, only change their policies if enough people feel and say that change is needed.

It is often difficult for an individual to look far enough beyond his or her own immediate concerns to consider matters on a planetary scale. This book has tried to show some ways in which worldwide ecological issues can directly affect us all.

Hopefully, it will have begun to show how the survival of rain forests, the conservation of wildlife, the use of safe energy sources, and the recycling of scarce resources are all part of an essential survival strategy for all life on Earth, including our own.

Ordinary people can help to set this survival plan in motion by taking part in conservation schemes, both locally and internationally. Each individual can help with conservation once they are aware of threats to our planet: by not using aerosol cans, by recycling packaging materials, or by growing wild flowers to encourage butterflies and insects.

Our living planet, viewed from space, is shown opposite for what it is—a beautiful, fragile film of life on the face of the globe. We will have to work hard and selflessly to make sure that it stays like this, to be enjoyed and lived in by future generations.

The planet Earth viewed from space

GLOSSARY

You may find it useful to know the meanings of some of these scientific words when reading the questions and answers in this book.

Adaptability The ability of a plant or animal to cope with all the challenges presented by its environment. An adaptable plant or animal is flexible enough to survive changes in its surroundings.

Algae The simplest sorts of flowerless green plants. They grow in fresh water and in the sea. The smallest algae are microscopic, and made of a single cell. Bigger ones form seaweeds.

Bacteria One-celled organisms—some of the simplest living things known on Earth. They may live independent lives or depend for existence on some other living creature. Many bacteria live in our bodies, and some of them cause disease. Bacteria are also helpful to humans. They improve the soil's fertility and are used in making cheese and other foods.

Broad-leaved The description given to a tree whose leaves have flat surfaces, that is, are not needle-shaped. Broad-leaved trees may be evergreen or deciduous.

Carbohydrate A chemical substance that is a complicated sort of sugar. Carbohydrates are one of the basic food materials needed by animals. Starches, and the substance cellulose, which are found in plants, are both types of carbohydrate.

Cell The basic unit of life. The simplest living plants and animals are made of one single cell. A complex creature such as a tree or a human contains millions of cells.

Colony A group of animals or plants that live together. In a colony the living things are of the same sort, for example, gulls or puffins.

Community A group of different sorts of plants and animals that live together, sharing the same environment. Each living thing in a community has an effect on its neighbors. Members of a community may depend on each other for life.

Deciduous Describes a tree or shrub that loses all its leaves at one season of the year.

Environment The surroundings in which an animal or plant lives.

Evergreen Describes a plant (usually a tree or shrub) that does not lose all its leaves at one season of the year.

Fungus Simple organisms that are not green, and therefore cannot trap energy from the sun. Fungi get their food by living on other organisms as parasites, or by dissolving and absorbing nutrients from the dead bodies of animals or plants.

Gene The part of the cell of an animal or plant responsible for passing on characteristics from one generation to another. The genes of an organism are responsible for the way it looks, works and behaves.

Germination The activity in which a seed develops into a grown plant.

Habitat The place where an animal or plant lives.

Invertebrate An animal without a backbone. Invertebrates are often described as soft-bodied. In some of them, however, the body is covered by a hard protective case, as in a crab or a cockroach.

Larva A young form of a creature, which hatches from an egg. A larva looks quite different from the adult animal. A caterpillar is an example of a larva.

Litter The layer of decaying leaves and other material found on the floor of a forest or wood.

Microbe A microscopic organism, but especially a bacterium that causes disease.

Mineral salts Chemicals found in the water in soil, or in fresh or sea water. They are taken in and used by plants to help them survive and grow.

Niche The "lifestyle" of an animal or plant; the particular way it lives and where it lives. A creature is adapted to life in its own niche.

Parasite A creature that depends on another living organism (its host) for all essentials of life. Parasites may live on the outside or inside of their hosts.

Photosynthesis The chemical reaction in which green plants trap energy from the sun and use it to build complex molecules from the simple ingredients of water and carbon dioxide gas. It is this crucial reaction that makes plants the "primary producers" on our planet.

Plankton The minute plants and animals suspended in the ocean or in fresh water.

Pollen The minute grains made by the male parts of a flower. Pollen must reach the female parts of a flower so that fruits and seeds can form.

Protein A complex chemical used by plants and animals to grow and survive. Plants can make their own proteins from simple raw materials. Animals take in protein building blocks in their food. Meat, eggs, and seeds such as wheat are important sources of protein.

Specialized Describes the way in which animals or plants, or the parts of their bodies, have changed to cope with a particular way of life.

Species A type of plant or animal. Living things of the same species can mate together and produce young which can themselves have young. Animals or plants of different species cannot do this.

Stamen The part of a flower that produces pollen.

Starch A complicated sort of carbohydrate. Starches are found in plants, especially in their seeds, and are an important form of food for animals. Corn, for instance, contains a great deal of starch.

Subtropical Describes parts of the world that are not hot enough to be tropical.

Temperate Regions of the world moderate in their climate, with clearly defined summer and winter seasons.

Tropical The hot, wet regions of the world around the equator.

Tuber A swollen underground plant stem that stores food material. A tuber may grow into a new plant. A potato is an example.

Tundra A cold region in the northern part of the globe which has no tall trees.

Vertebrate Any animal with a backbone inside its body. Fishes, birds, and mammals are all vertebrates.

Index

Page numbers in **bold** type indicate illustrations

Acknowledgments

Artwork by
Richard Orr
Michael Woods
Vana Haggerty
Colin Newman

Maps by
Eugene Fleury

Photographs
11 E.A. Janes/NHPA; 22 *tl* Marc Romanelli/The Image Bank; 22 *r* Brian Rogers/Biofotos; 22 *bl* Edward Bower/ The Image Bank; 26 *t* John Shaw/NHPA; 26 *b* Ivan Polunin/NHPA; 27 *t* M.P.L. Fogden/Bruce Coleman; 27 *b* Dr John Dransfield; 29 John Payne/ICCE; 31 Hutchison Library; 33 Tom Willock/ Ardea; 37 The Bettmann Archive/BBC Hulton Picture Library; 43 Bryan & Cherry Alexander; 46 David Hughes/ Bruce Coleman; 49 *t* Lucian Niemeyer; 49 *b* Dan McCoy/ Rainbow; 53 Oxford Scientific Films; 57 Ron & Valerie Taylor/Ardea; 59 *t* Peter Parks/Oxford Scientific Films; 59 *b* Robert Hessler/Seaphot; 62/3 Georg Gerster/The John Hillelson Agency; 64 M.P.L. Fogden/Bruce Coleman; 71 Georg Gerster/The John Hillelson Agency; 74/5 John Shaw/ Bruce Coleman; 77 Pam Isherwood/Format; 81 Gamma/ Frank Spooner Pictures; 82/3 Burt Glinn/Magnum Photos; 83 I.R. Beames/Ardea; 85 F. Hartmann/Frank Lane Picture Agency; 88/9 Chris Davies/Network; 91 NASA/Bruce Coleman.